The Importance of
Being Interested

'A delightful and scintillating hymn to science. Resolutely a non-scientist, Robin Ince discovers with awe that when science addresses the "big problems" and destroys familiar beliefs, it does not leave us in a cold, meaningless and dehumanized world, but in one that is colourful, human, and full of intensity and wonder.'

Professor Carlo Rovelli

About the Author

Robin Ince is the co-creator and presenter of the BBC Radio 4 show *The Infinite Monkey Cage*, which has won multiple awards, including the Sony Gold and Rose d'Or. In 2019 he played to over a quarter of a million people with Brian Cox on their world tour which has put them in the Guinness Book of Records for the most tickets sold for a science show. He is author of *I'm a Joke and So Are You* and also won Celebrity Mastermind but forgot that calcium was the dominant element of chalk. He is currently trying to invent an effective satnav for people who believe the world is flat.

The Importance of Being Interested

Adventures in Scientific Curiosity

Robin Ince

Atlantic Books
London

First published in hardback and trade paperback in Great Britain in 2021
by Atlantic Books, an imprint of Atlantic Books Ltd.

1 2 3 4 5 6 7 8 9

A CIP catalogue record for this book is available from the British Library.

Hardback ISBN: 978 1 78649 262 3
Trade Paperback ISBN: 978 1 83895 429 1
E-book ISBN: 978 1 78649 263 0

Chapter header illustrations by Mecob, based on images from Shutterstock.
Internal illustrations: p36 © Getty Images; p91 © Robin Ince; p103
© Wikimedia images; p138 © European Space Agency; p168, p355
© NASA; p353 © ADAGP, Paris and DACS, London 2021.

Printed in Great Britain by Bell and Bain Ltd, Glasgow
Atlantic Books
An imprint of Atlantic Books Ltd
Ormond House
26–27 Boswell Street
London
WC1N 3JZ

www.atlantic-books.co.uk

*To all the librarians and teachers who have
encouraged our curiosity,*

*And to my sister,
sometimes Camilla, sometimes Janey,
depending on whether my parents
thought she was being good or bad,
who has maintained her fascination with the world
while often keeping it turning for other people.**

---

* This now puts huge pressure on me to dedicate my next book to my other sister, Sarah. I will try to write one as soon as possible.

Contents

Foreword

Richard Feynman once wrote that scientists' most valuable transferable skill is a deep and intimate experience with doubt. It's difficult to motivate yourself to spend a life in research if you believe you know everything, and even the most self-confident research scientist will ultimately be humbled by their encounters with Nature. This is the best argument I know for maintaining at least a small component of science throughout every citizen's education. I once half-jokingly wrote that the PPE course at Oxford, studied in the loosest sense of the word by many a cabinet minister, should be rebranded PPES; perhaps brushing up against Nature occasionally would moderate their certainty. After all, as Feynman also pointed out, democracy itself rests on the acceptance that we don't really know how to run a society; that's why we change our politicians every four or five years. If you think you *know* how to run a country, if you think your policies are absolutely right and the other lot are absolutely wrong, you are not a democrat.

Robert Oppenheimer came to similar conclusions in his 1953 BBC Reith Lectures. Nature forces us to hold seemingly contradictory ideas in our heads in order to understand what we observe. A thing as simple as an electron is sometimes best thought of as a wavy, extended object and sometimes as a point-like speck. Crucially it is neither, but both pictures are necessary components of our understanding. Similarly, society may appear to be riven by tensions between the competing human desires for individual freedom and collective responsibility: but riven is the wrong word, because both desires are present in every individual and both are therefore necessarily present in society. The democratic process gently swings the pendulum one way and the other, and the swing is both the manifestation and guarantor of our freedom.

Your freedom, then, is protected by your acceptance that you might be wrong, and science is a sure-fire way of forcing you to practise being wrong. Which brings me neatly to my friend and colleague Robin Ince. He describes his role on *The Infinite Monkey Cage* as that of professional idiot. He means this in a self-deprecating way; indeed, he has elevated heartfelt self-deprecation to something of an art form. From his wardrobe to his gait, he radiates uncertainty. But, as I have argued, there is no more valuable skill. There are two categories of idiot: the curious idiot – a category that includes all scientists – and the idiot – a category that includes all who are certain. Robin is a category one idiot, and that's why he's an engaging and wise guide.

Robin's thirst for knowledge is unquenchable, and in these pages he engages in debates and conversations with a dazzling cast of great minds in search of a little enlightenment; not absolute enlightenment, because that's not on offer. We do not understand the human mind, we do not know what it means to live a finite life in an infinite universe, and we do not know whether or not there is a God. If answers exist, they reside in unknown terrain, and that's what makes them interesting. It's important to explore that terrain with humility and an open mind. It's important to be interested.

Brian Cox
July 2021

The Stars Your Destination

In studying how the world works we are studying how
God works, and thereby learning what God is. In that spirit,
we can interpret the search for knowledge as a form of
worship and our discoveries as a revelation.

Frank Wilczek

The moment I put my hand in my school-blazer pocket and
found it full of frog entrails, I already knew science was
not for me.

As a young child, I loved science. Primary-school science
classes were full of excitement, whether it was interrogating
leaves or watching Robert Calvert see blood and then faint and
smash in his front teeth. In secondary school, though, this joy
evaporated. I think many people lose their interest in science at
secondary school, and I was one of them. This is where science
became serious, but also where it became joyless. This is where

the equations and explanations seemed detached from my own experience. Whatever science was, it was not *lived* experience. It was as if scientists only thought in sums. They didn't daydream and play. Each day they opened their box of numbers and symbols and moved them about until they were satisfied: 'I've carried the two and now I am satisfied that I have a testable wave function.'

That was predominantly how I felt about science at school. My curiosity about the world never went away – I just ignored it. I'd be thrilled to see an afternoon of art on the timetable, whilst to know that double physics was coming was to foresee time moving slower than Einstein could ever imagine. The division between the arty and the science subjects seemed jagged and high. The two cultures were clearly growing in separate Petri dishes.

Our physics teacher, Mr George, was a clever man, but the sort of person who didn't understand people who didn't understand. He was also easy to antagonize, and so the bullying, disruptive boys would hide his pens and see him explode apoplectically. They would giggle as he held back tears. It was a horrible sight, made worse after I read *The Lord of the Flies*, as I now knew exactly which of the boys in my class would be the ones shattering my glasses and throwing me off the cliff onto the rocks below. Putting frog entrails in my school-blazer pocket was simply the beginning of what might be possible.

My chemistry teacher had clearly lost interest many years ago. Biology had a little more pizzazz about it. There was the relative excitement of watching incubated locusts shed their skins. Then there was Mr Rouan, head of the department, who still had an enthusiasm about his subject and was always ready with a slightly inappropriate biological quip. I remember a nervous colleague of his embarking on sex-education hour, demonstrating how to use a condom with a broom handle playing the part of a priapic penis. The teacher's shaking hand lost its grip on the broom and it fell to the floor, just as Mr Rouan passed the doorway and made a rude joke. We all laughed loudly as we pretended to understand.

But mostly I remember the mind-numbing effect of an afternoon double-physics class, head on the desk, drooling over my exercise books as the teacher plugged away at speed and mass equations. A dismal result in a physics exam, and an increasing sense that science was detached from the real world, finally put the nail into any delight in science. The tables where the science boffins sat had a strange aura of unknowable exclusivity. It was clear to me they had different brains. They understood this stuff. They were the Midwich cuckoos, otherworldly and somewhat threatening.

Now I wonder how on earth I ended up co-hosting *The Infinite Monkey Cage*, the long-running radio show and podcast. My early adult life was spent building a career in comedy, but sometime in my mid-twenties I bought a book about quantum physics. I didn't really understand it, but I realized that what I wasn't understanding was very exciting. I carried on reading books I didn't understand, with only small glimmers of occasional comprehension. I started to refer to some of the ideas in comedy routines, then I began to base whole shows around them, and from there it only seemed right to ask whether real scientists could join me onstage to make sense of my confused ramblings. This eventually led to *The Infinite Monkey Cage* with the physicist Brian Cox. Then I started joining him on tour, where he would explain high cones and holographic principles, and I would ask the dumb questions I realized I'd always wanted to know the answers to, and that others might be afraid to ask because they feared they might look stupid.

At the time of writing there have been 150 episodes of *The Infinite Monkey Cage* and it has covered everything from the theory of relativity and the Higgs boson, to how science proves that it is best to eat a pear with a golden spoon, and how to speak fluent chimpanzee. I have spoken to Nobel Prize-winning geneticists, *Apollo* astronauts, undersea explorers and one wizard. Such work has often meant that I'm in the fortunate position of usually being the stupidest person in the room. It's not always good for the ego, but it is very good for my education. W. H. Auden wrote that when he was in the company of scientists, he felt 'like a shabby curate who has strayed by mistake into a drawing room full of dukes'.[1] I am pretty happy to be the shabby curate. I've got the wardrobe full of cardigans, and I come from a long line of vicars, so I have that ecumenical gene.*

The way the guests on the show explain and talk about science – the way they make it relevant to everything about my daily life, my existence here on this planet, the past, the present and the future – has rekindled my enthusiasm and widened my curiosity for a subject that died a death on the Bunsen burners of my youth.

I've realized, though, that whilst much can be made of what we have gained from scientific knowledge, for all the enthusiasm

* Yes, yes, yes, I am aware that they have not really isolated the vicar gene. Nature and nurture vie their way to propel you to the pulpit.

4

and passion of the scientists I speak to on the show, much can also be made of what has been lost. For some people, while science has given us the power to do things, to create, to control and to live longer lives, it has also delivered a longer life that now seems meaningless. Some say that science has robbed us of our gods, our exceptionalism, our centrality, our free will, and leaves us lost and alone in a vast universe. I can understand why some may feel this way, if they merely glance at the knowledge we have gained from science, but I believe that the deeper you explore science, the more our new knowledge creates rich stories, new enchantments and, rather than leaving us alone, connects us to everything. But I am getting ahead of myself; let's go back to the Garden of Eden and start again.

Did it all go wrong when we ate from the fruit of the Tree of Knowledge? Before then we were living in blissful ignorance, but then curiosity kicked in, and with that curiosity came questions and doubt. Some suppose that the universe was enchanting when we knew nothing of it. The lights in the sky twinkled because they were attached to a celestial sphere or shone through the holes of a heavenly curtain; but then they became nuclear reactors, shining as hydrogen became helium, some of them destined to collapse into something so terrible and mighty that even light could not escape. Your downfall begins the first time your face creases into a frown and you say, 'But why?'

'Don't ask why, just get on with it!'

The best way forward for some people is to stay exactly where they are. 'Ours is not to reason why...' Perhaps there's a fear that if we pull back the curtain, we might be disappointed by what we see. 'Pay no attention to that man behind the curtain. The great Oz has spoken,' says the charlatan in *The Wizard of Oz*, although the result of the revelation that the great Oz is simply a man with a megaphone and a smoke machine actually leads to satisfying answers for all on such issues as the heart and the brain.

Curiosity, in particular scientific curiosity, is dangerous to the powerful. Power often rests on certainties, and the scientific method encourages active doubt. There are warnings to the curious. As Oedipus found out, 'How terrible is wisdom when it brings no profit to the wise.'*

Questions can be seen as impertinent and dangerous. You may lose everything you hold dear – your god, your afterlife, your free will, your feeling of superiority, your mind... even your entire reality. Knowledge can be framed as loss rather than gain. It is a quandary I have been dealing with for the last fifteen years.

I often ask myself how scientists came to be doing what they are doing. Why weren't they bored to tears in their science lessons? Do their brains work differently? What do they see when they are explaining quantum indeterminacy? Are they born with scientific brains? Is the ability to understand supernovae or charm quarks somehow hard-wired? This was one of my first anxieties when I started making science shows. Was I allowed to think on such things? Did I have permission even to ponder these subjects, without qualifications? Scientific ideas can seem so daunting that they may feel both forbidding and forbidden. Any question from a novice like me surely has a high probability of being a stupid question.

* Obviously I would like you to think that I am quoting from Sophocles, though I'm really quoting from the Mickey Rourke film *Angel Heart*.

It can be easy to believe that scientific ability is built into us by a quirk of nature – our genetics. If you find science hard, it is because your father found it hard, and your father's father found it hard, and you have inherited the 'not understanding cosmology' gene. This was how it seemed to me. I struggled with learning science at secondary school and ended up believing that I didn't have the correct configurations in my brain to check into Hilbert's Hotel or diagnose Schrödinger's cat. I don't think I'm alone in believing this; but that so many people should presume they are unfit for science perhaps suggests there is something wrong both with how we learn science and with what we believe it to be.

Even Carlo Rovelli, who is a founder of the loop quantum gravity theory and a writer of very beautiful books on physics, struggled with the tedium of some of his science education, but he was able to see beyond it. As he writes in his book *Helgoland*, 'What attracted me to physics was that beyond the deadly boredom of the subject taught in high school, behind all the stupidity of all those exercises with springs, levers and rolling balls, there was a genuine curiosity to understand the nature of reality.' Fortunately I have found a way back into feeling a fascination about science, though I can assure you that I will not be contributing anything of any significance to loop quantum gravity theory.

I have totted up daily the pros and cons of confronting my ignorance, and I am pretty sure the advantages outweigh the disadvantages. Sure, it means I now live in a meaningless universe, by the looks of things, but existential philosophy was eager to tell me that, before astrophysics ever got involved. If you want to feel frighteningly alone in the universe, sit on a railway-station platform in midwinter, waiting for a train that increasingly looks as if it will never come, and read Jean-Paul Sartre: 'Every existing thing is born without reason, prolongs itself out of weakness, and dies by chance.' And then he doubles down on that with: 'It is meaningless that we are born, it is meaningless that we die.'* These are the sorts of aphorisms that would lose you your job in the fortune-cookie factory.

Professor Brian Cox is very fond of the words of John Updike: 'Astronomy is what we have now instead of theology. The terrors are less, but the comforts are nil.' But does this mean there are no comforts or consolations from science, bar perhaps the temporary consolation of medical ingenuity, spaceships and instant-whip desserts?

Sometimes it can be hard to start the day with a spring in your step when you have been made aware that you are merely a perturbation in the universe's wave function. The last few centuries have seen our uniqueness being whittled away – we are no longer at the centre of the universe, no longer a special creature separate from those grubby, ball-licking, poop-flinging animals. Like all of the rest of them, we are just a quantum fluctuation, although a quantum fluctuation that combs its hair and plays Scrabble.

Physicists usually seem the least bothered by such a demotion. I think it is because they see things either at an atomic level or

---

* When Jean-Paul Sartre was a little boy, he had lovely blond curly hair and his mother thought he was a beautiful angel. Unfortunately, while she was out, his grandfather insisted that his 'girlish' locks were snipped off. Without his curls, his mother thought he looked monstrous and, upon seeing him, ran to her room. That sort of thing can lead to a boy being a pioneer of existentialism.

8

wave-function level, and find the recycling of these patterns and subatomic particles satisfying enough. Biologists seem a little more concerned, perhaps because they observe things at a more molecular level and smell the organic decay. Chemists generally don't have time for either position, as they are too occupied with wondering why people don't talk about chemistry enough. Chemistry is the middle child. First there was physics, the older sibling; finally there was biology, the spoilt child; and in between, chemistry came into the universe: essential, but often overlooked.

The physicists seem to have got so used to the indifference of the universe that they forget it might be news to other people, and they forget the need to break it to us gently. This can lead to nihilistic flourishes at public lectures and debates, which can be deeply disturbing. They bandy about our inconsequentiality and expect us to sit obediently, taking it all in. We are *just* a bunch of atoms. *Everything* is just a bunch of atoms – Chartres Cathedral, the Grand Canyon, a blue whale, Jupiter. Cancel your travel plans; you can simply stare at all the atoms that you have at home; they may well form something magnificent one day, so enjoy them while they take the shape of your desk-tidy or pan-scourer. It is like returning from a world tour and dismissing the Great Wall of China as 'just a load of bricks'.

Our experience and sensations are all down to nothing but firing neurons. You are merely calcium ions, firing away. Even our selfhood may be an illusion, as is our autonomy. We're on a humdrum planet. We're in a corner of the galaxy that is unexciting. Our galaxy is mediocre.

You can see why scientists don't always make the best motivational speakers, and why first dates can be tricky, because not everyone wants to know the number of bacteria living on the surface of their skin, before the starter. Yet again, though, it is the philosophers who most firmly bop us on the nose. The philosopher Bertrand Russell wrote, 'The universe has crawled by slow stages to a somewhat pitiful result on this earth and is going to crawl by still more pitiful stages to a condition of universal death.'[2]

Some may see such statements as cosmological honesty, but, like a work colleague who says, 'A lot of people in the office think you smell of Camembert stuck in a burnt-out clutch, and that your new haircut makes you look like Mao Tse-tung – just thought you should know',[*] such plain speaking can be upsetting and a source of despair and depression, even if it might be true. 'I speak as I find, and I find the universe to be indifferent, and destined to end leaving no trace of human creativity or indeed any knowledge, love or beauty at all. And how was your day at the office?' If that has been your only experience of engaging with science, then I can see why many people decide not to return to it, but in my view that is in fact why it is worth sticking with. First may come disillusion, but then comes re-enchantment. Some discoveries hit harder than others, but if you can get through that existential pain barrier, there are things on the other side (I am not one for running marathons, so I work out at the library instead).

With our loss of exceptionalism, there is also a gain of connection, and these connections can be found across scientific disciplines. You are not alone up on your Olympian heights; you

* I should make it clear this does not come from personal experience. I have never had a haircut that made me look like Mao Tse-tung, though I did have some glasses that made me look a bit like Ernest Bevin.

are joined up to everything, and the loneliness of uniqueness is replaced with a new cosmological camaraderie. A scientist's pessimistic realism is often the most coherent and quotable way of accessing their work, but after the humdrum, after the 'cold and indifferent', comes a big BUT...

I believe that within much of what can seem to be negative or pessimistic about our universe, there are many possible theories that can propel us, that drive us to find our own meaning and consolations – and hopefully that is what this book is about. You don't have to depart from reality to find happiness and purpose. There are meanings in all this fragility; there is wonder and delight in all these doubts. Detaching yourself from certainty does not mean you must feel lost and bereft. The problem with ultimate truths and utter certainties is that they can get in the way of your adventures in ideas and can possibly block paths altogether. Accepting that the inevitability of life must be attached to the inevitability of death should sharpen the senses and the need to experience. The realization that to love is also to commit to loss is what magnifies that love.

Facing the realities of what scientific endeavour can tell us about the universe, and ourselves, can seem like facing up to the loss of things we have relied upon or held dear, but with the losses come gains that outweigh them, even if they are not always immediately apparent. I think the realization that there is no grand meaning to us – that we are not born with meaning stamped on us, but must strive for meaning, in all its tentativeness and potential fragility – makes it far more vital.

And yet for all these grand philosophical ideas about meaning, returning to actual engagement with science and scientists can be a bumpy road. Sometimes, many years after last burning your fringe on a Bunsen burner, reopening a science book can be a disappointing and frustrating experience. You start to read a book about quantum theory because, bizarrely, someone said it would help you understand how things may be alive and dead all at once, which sounds amazing; and you'd also really like to understand that Christopher Nolan film you have watched three times now. Dismayingly, though, as you plough through the book, it gets more and more complex, you fail to understand superpositions and entanglement, nothing seems to relate to what you really wanted to know; and you end up throwing your arms up in the air and presuming that you're stupid and, once again, simply do not have the brain required. Sometimes the voice in my head shouting, 'YOU DON'T UNDERSTAND THIS!' is so loud that I can't even hear the sentences I am reading.

At times the words on the page can seem to have a life of their own, independent of the reader. I have often found myself rattling my skull, desperate to work out where all the information that I have just read has gone. Was each sentence like a neutrino, passing through my eyes and skull without ever interacting with my brain?* Even if it does begin to make an impression, some of what you start to understand is aggressively counter-intuitive. Cosmological common sense seems to be in limited supply. For instance, it took me a very long time to get my head around the idea that there are

---

* Neutrinos are a good example of being confounded. When I first read of how many pass through us constantly, I could not understand how something can exist with mass, but mass that is so small it passes straight through me. For the first few days after finding out about them, I am sure my body had an increased sensitivity and could feel at least a few of those 100 trillion neutrinos that were passing through me.

200 billion stars in our own galaxy alone.* Every time I said it out loud, I presumed I would be openly mocked.

Then I found out that many astronomers believed there were more galaxies in the universe than there were stars in our galaxy. Then I was told that the size of the universe could be infinite, which means there is someone else across the universe who has just read 'the size of the universe could be infinite, which means there is someone else across the universe who has just read "the size of the universe is infinite"' – and they have a head and life exactly like mine, or I have one exactly like them: different atoms, same life. That goes for you, too. And you.

That there is nothing special about me means that there are an infinite number of *me*'s. There are an infinite number of *you*'s, too. Then I read that we have to say '*our* universe', because we are probably one of many universes. I bump into a quantum physicist who is keen to tell me about 'many-worlds interpretation', where everything that can happen does happen and, at the point of each potential event dividing, more worlds are created to allow all possible outcomes to occur. Now you have a multi-verse of many worlds.

Another cosmologist butts in and tells me about the 'holographic principle' of black-hole thermodynamics, which suggests that all physical objects – including ourselves – are actually two-dimensional projections from somewhere else. On top of all that, I am still trying to get my head around the idea that my head, and everything else in the vast known universe, used to fit on the

---

* I eventually found out how many stars there were in the universe: (approximately) 70 sextillion or, rather, 70000000000000000000000. Then I had to listen to whale song in a darkened room for a few days.

end of something smaller than the prick of a pin. Actually, even smaller than that – a sort of nothing-whatsoever size. Everything I have ever imagined was contained in something of infinite density, but no mass.

How could that be? I find it hard to close a suitcase if I try to put a spare pair of shoes in it, let alone a spare plant, hat stand or galaxy. It can all seem like the fevered imagination of a speed-pepped science-fiction author fearful of missing his deadline for *Astounding Stories*. You can see why people might shy away from science – never mind the numbers. It feels utterly absurd.

But if the universe was easy to understand, it would be a very boring place. When you see the professional public scientists broadcasting, they often seem sure, certain and infinitely polymathic. This is why some of the most important moments to watch out for may be when you see the scientists perplexed. When faced with a question to which the scientist's reply is 'I don't know', we can feel immensely relieved, but sometimes this can be followed by excitement. 'Now you ask, let's see if we can work it out!'

I remember standing with a physicist in front of an audience of 4,000, trying to work out how a Slinky moves downstairs. We came up with interesting answers, both of which were wrong, but even getting to the point of error was fun, and hopefully many people went home from that event, found their old Slinky and started their own research work. It hasn't just been Slinkies, though; it's been thinking about the edge of the universe, about the possibility that we are the only intelligent life in the universe (and that may well not be intelligent enough) and about the Sun swelling into a red giant. Sometimes such pondering is playful, and sometimes the air of doom becomes sweltering. And that's really been my inspiration for this book.

When I first started re-engaging with scientific ideas, it was easy to get lost. It still is. It is a big universe and there are many ideas and theories about it, but the anxiety of not knowing where I am is not as jagged and forbidding as it once was. Scientific progress and development can fill people with confusion and fear, and it can challenge their most deeply held beliefs and connections, but I have lived to tell the tale and to want to learn even more stories. My mind has been repeatedly blown by the images and ideas offered by scientific thought and enquiry, and I am glad. I am getting used to doubt, and I am inspired by the seemingly inexplicable. I don't need a quick fix any more. A little knowledge is only a dangerous thing if you think it is enough knowledge.

Brian Cox once wrote, 'A little existentialism never did anyone any harm.' But when I asked him about this, he admitted that he didn't think he had ever experienced any existentialism; he simply imagined it might be useful for people who did. I am the anxious one in our partnership. Our temperaments are a cliché of art versus science. I am the fraught, antsy bag of nerves, while he coolly wanders towards the certainty of his own demise, safe in the knowledge that his atoms will survive, even if he doesn't. Brian wouldn't rail against the dying of the light; he would simply capture the light in an equation.

I believe that almost any loss that comes from the scientific adventure carries with it great gains, too – not merely pragmatic, but also enchanting and transcendent ones. Whatever idea seems to rob you usually contains a reward as well.

I wrote much of this book during the first lockdown of the Covid-19 pandemic. A positive outcome of the pandemic for me, in this regard, was that a number of people who would never usually have been available were kicking their heels at home, so bored that they talked to me. When I look at the wish list that I drew up at the beginning of 2020, the only person I failed to talk to was the film director David Cronenberg and, to be honest, it wasn't so much that I needed to talk to him for the book; I just love *The Fly** and I thought I could add a few paragraphs to the chapter on biology, dealing with the current genetic understanding of human–fly hybrids, which can be the outcome of drunken use of a teleport. As I talked to all the different contributors to this book, I have found the picture of the universe around me changing frequently. I think one of the purposes of bold human endeavours – whether scientific, philosophical or artistic – is to change how we see what we see, and possibly change ourselves with that.

Changing your mind is not always easy. These days, particularly across politics and on social media, it is easy to find people who would rather be aggressively certain than tentatively contrite or in doubt. Many seem to believe that it is better to be solidly wrong than wavering towards being better informed. But a common theme with the many people I spoke to was the need for inquisitive humility rather than righteous brutality, if we are to progress. And with that humility comes the need to

---

* And *Scanners* and *Eastern Promises*, and *The Dead Zone* and *Spider* and...

16

interrogate yourself as much as you interrogate other people and to ask, 'Why do I believe what I believe? What foundations am I standing on? And why do I favour them?'

We need to know who we are, in order to find out who we can be; to refute pointless barbarism and squabbles and build a world on common ground and collective understanding. I also believe that the more we confront the meaninglessness of the universe, the greater our ability to create our own meaning. I have tried to deal with the areas of understanding that have robbed us of our more comforting myths, but this has not robbed us of stories. The universe is still made of stories, and they have the comfort that they may well be true, too.

So I hope this book can do that for you. I hope you find something in it to enrich your picture of the world, or that you find some new enchantment on what might have seemed to be barren land. From my very first interview with the astronaut Chris Hadfield, a human who has watched the world turn beneath him, to my last interview with Carla MacKinnon, an artist whose experience of sleep paralysis means she has felt night-hags squatting on her chest, the shared underlying message of so many people has been that the more we explore and the more we learn, the better our questions become, the greater the adventure and the more connected we are, whether it is to supernovae or to octopuses. Life becomes easier to live when you start to understand it, when you don't ignore the questions, when you don't try and paper over your confusions, but open up to them.

By becoming acquainted with scientific curiosity, by learning and understanding from it, I believe we can be re-humanized rather than dehumanized. Perhaps we can be as beguiled by reality as we can be beguiled by myth, and can find room for both.

I am extremely timid, sensitive, impressionable, and with a great sense of mortality, lazy. Yet inside me there is another self completely unmoved by all of this, full of power and light.

CECIL COLLINS

Not explaining science seems to me perverse. When you're in love, you want to tell the world.

CARL SAGAN

CHAPTER 1

Scepticism – From the Maelstrom of Knowledge into the Labyrinth of Doubt

Doubt is not a pleasant condition, but certainty is absurd.

Voltaire

I would love to be certain, to be sure of something, but my anxiety makes that unlikely. Perhaps my perpetual nervous doubt has had some advantages for me, as it has meant that whilst science has encouraged a deepening of my scepticism about the world, I have not had to painfully sever myself from any strongly held, dogmatic beliefs.

In his book *Science and Hypothesis* the French mathematician and physicist Henri Poincaré wrote, 'To doubt everything and to believe everything are two equally convenient solutions; each saves us from thinking.' Finding the Goldilocks portion of doubt

– the doubt that is 'just right' – can be tricky. The author Robert Anton Wilson, who co-wrote *The Illuminatus Trilogy*, a wonderfully playful and adventurous romp through all the conspiracies there have ever been, encouraged universal agnosticism. This was not merely agnosticism about the gods; this was agnosticism for all your beliefs.

Good science is never certain. Though it might be sure it is providing the best answer for the time being, you always need to be prepared to loosen your grip. This is not only true of science, it should be true of all beliefs, but it is not easy. We are tribal and we like to feel that we belong. To belong often meets to be united by your beliefs and, even more so, united by hatred of those who contradict those beliefs.

We are all victims of our cognitive dissonance. Our desire to hold on to beliefs about the world means that people often follow all manner of circuitous routes of thought in order to preserve their beliefs, even when they become increasingly preposterous. With the dominance of social media, there is now a perpetually pumping plumbing system that sprays out opinions twenty-four hours a day. We are never more than two seconds away from an utterly bizarre opinion that is rabidly held.

Sitting in rooms with physicists, I sometimes observe their perplexed faces as they try to understand the erratic and destructive beliefs and decision-making of other humans. 'Haven't they seen the statistics and the graphs?' they wonder. For some, this is why they were drawn to physics. Even in a probabilistic universe, the paths of electrons are easier to predict than the actions of other human beings. Evidence can often play a very minor part in *why* we believe what we believe, so when the physicist tries to refute your ideology with years of carefully accumulated evidence, you can shrug it off and return to your invisible leprechaun farm.

I can also find myself frustrated by scientists. It seems to me there are some who consider that they have such control over their own minds that the pure evidence they have discovered means they would never fall victim to any cognitive dissonance. I have also grown to dislike those sneering T-shirt slogans or memes that say, 'Science doesn't care about your opinions', as if by being a scientist, their feelings and biases are always usurped by their superior brains, which can manage to neatly box their emotions, only occasionally releasing them for birthdays, funerals and screenings of *The Shawshank Redemption*.

It goes both ways. When scientists are particularly attached to an idea, evidence may not be enough; and again, though they might deny this, emotion may play its part.

Fred Hoyle was a brilliant scientist. With US physicist Willy Fowler, he demonstrated that all the elements of our world originated from inside stars and were then projected through the universe via stellar explosions. The romance of us being made from star-stuff was confirmed by his work. It provided us with a beautiful story, offering us another sense of connection with the whole universe. Pondering on the journey that the atoms that make up you and me have made throughout history is meditation time well spent. Fred Hoyle is probably best known, though, as the astronomer who, despite increasing evidence, would not accept the Big Bang theory. He continued to prefer his steady-state theory. Astrophysicist Chris Lintott thinks that what lay at the heart of Fred Hoyle's thinking wasn't a scientific dispute; it was his hope and desire that the universe would go on for ever. After all, when the universe ends, physics ends, too. 'Hoyle realized very early that if you have a universe that's changing, that implies not just that there's a beginning but that there's an end, even if it expands for ever. And so his steady-state theory was a way of getting away with that, because you're continually producing new raw material from which you can keep making stars and planets and astronomers,' Chris explains. When considering scepticism, it is important to realize that the clear thinking of scientists may also be polluted with emotional attachment and egotism.

With curiosity, though, comes doubt; and if your doubt remains active, you may become marked as a sceptic – something often confused with being a cynic. The sceptic, like the atheist, can be seen as a killjoy. There you are, having all the fun of believing that the Earth is flat, or shunning a potentially life-saving vaccine because a Hollywood celebrity made a five-minute YouTube film that was informed by someone else's five-minute YouTube film, which was in turn informed by a YouTube film by a vitamin-pill salesman (and the infinite regress of misinformation goes on), and some sceptic comes along and suggests that it might all be a bit more complicated than that. The importance of doubt working in tandem with science is potentially life-saving. Some may say at this point, 'But isn't rejecting the Earth being a sphere, and rejecting vaccines, scepticism?' And I would suggest that you try arguing with the people who hold these opinions. Their beliefs require a rejection of a great deal of evidence, often based on suspicion or paranoia and the unproven ideas of secret cabals, not really offering testable alternative realities or ideas.

There is a reason why fundamentalist religious and political systems often ban books. It's because they can be full of ideas

and possibilities that can reframe the world in a way counter
to that set out by those systems. Books are pliers for the barbed
wire with which a dictator wishes to encircle someone's mind.
Once, in Canberra, I spoke to a man who had been brought up
in a fundamentalist Christian commune. While other teenagers
in Australia may have been hiding creased pornography under
their mattress, he had a book of essays by the philosopher A. C.
Grayling. This was the key to his rebellion.

A book brought me to sceptical thinking, too, though I did not
have to hide it under the mattress, which was fortunate, as I was
sleeping on a futon with a very thin mattress and the sharpness
of a book corner could have caused spinal damage. In *Sceptical
Essays*, Bertrand Russell wrote that scepticism will diminish the
incomes of clairvoyants, bookmakers and bishops. He tells the
story of Pyrrho, the first Greek sceptic philosopher, who considered that 'we never know enough to be sure that one course of
action is wiser than another'. When Pyrrho saw his philosophy
teacher with his head stuck in a ditch, he didn't pull him out, as
he didn't think he had enough information to be certain that his
teacher didn't want to have his head stuck in a ditch. After others
pulled the teacher's head out, his teacher congratulated Pyrrho
for correctly interpreting his philosophy, even if it did lead to him
having his head stuck in a ditch for longer than it needed to be.
(We are never told how he got his head stuck in a ditch in the first
place, and I am not sure I would hold someone's teachings in such
high regard if they were the sort of person who found themselves
in such a position. I am also not sure he would have congratulated Pyrrho quite so heartily if no one else had pulled his head
out of the ditch. By the third day, say, even the most stoical of

stoics could have become fed up of having his head stuck in a ditch, especially if goats were beginning to sniff around.)

'They swim. The mark of Satan is upon them. They must hang'

My interest in scepticism grew before my interest in science was rekindled. It really took hold at the witch trials, or at least the site of witch trials, both historical and fictional.

It was the story of the small English town of Lavenham that sparked it all off for me. Its connection to witch trials made it a good place to start questioning *why* we believe what we believe. Here was one of the locations where the vagaries of nature – whether crop failure or erectile dysfunction – had led to women being burnt at the stake if they failed to do the proper thing and demonstrate their propriety by drowning in a river.* ** Women from Lavenham were tried as witches, but the most famous Lavenham witch trials were fictional, in the film-cult classic *Witchfinder General*, the final film of the far-too-short

---

* As this is a chapter on scepticism, I should be clear and tell you that only one woman accused of being a witch was executed in Lavenham. Her name was Anne Randall. According to the court proceedings, she sent her Imp Hangman to kill a horse of one William Baldwins of Thorpe. She confessed that after begging for alms and being rebuffed, her Imp Hangman appeared to her and asked her what he should do, and she bade him go and kill a man's hog and that 'she being angry with one Mr Coppinger of Lavenham, she sent her Imp Jacob to carry away bushes, which he had caused to be laid to fence his fences'. Apparently the hedgerow where she was executed still shows signs of being cursed.

** Oh, and after discussions with a variety of authors and researchers, it is also probable than Anne Randall was not burnt at the stake, but was hanged, though this is not certain. The general punishment in England for witchcraft was hanging – burning at the stake was more of a film conceit. The problem with writing a chapter on scepticism is that you have to be extra-thorough.

life and career of *enfant terrible* film director Michael Reeves,[*] with Vincent Price playing Matthew Hopkins in the title role.

After wandering around the Lavenham square where Price's Hopkins set fire to Maggie Kimberly, I went into an antique shop, looking for a Toby jug of interest or a framed cigarette card of Boris Karloff. Instead I found a second-hand copy of James Randi's *Psychic Investigator*, the book of the TV series in which the conjuror and escapologist tests the claims of spiritualists, dowsers, telepaths and suchlike.

My first attraction to these stories was as low-hanging fruit for stand-up comedy routines. I was particularly keen on the toe-curling tales of floundering psychic mediums struggling for any sort of ghost that might connect with their audience. My favourite was the medium who stood onstage and announced that the ghost who had sat down beside him liked cheese-and-pickle sandwiches – one of Britain's most popular sandwiches, especially among the generation that were most likely to have died recently, and likely to be popular with his living audience too. Somehow, in a room of more than 500 people, none had a deceased relative who was remembered for being fond of this highly popular sandwich – surely a statistical aberration. Labouring over the details of the snack, as if the audience might have forgotten what a sandwich was, the increasingly flustered and irritated psychic still found no takers. Finally a sheepish

---

[*] I worked with one of the stars years later, Nicky Henson. He told me some delightful stories of the antagonism between the vigorous young director and the veteran actor. In a fit of pique after a directorial suggestion, the fifty-seven-year-old Vincent Price said, 'I've made eighty-seven films, what have you done?' To which the twenty-three-year-old Reeve replied, 'I've made three good ones.' A fine comeback, though not entirely factually accurate.

man at the back recalled that his father loved cheese, but that he was not allowed to eat it. This was the spirit connection that the medium now said he was looking for. Due to poor communication between the living and the dead, 'lactose-intolerant' was translated as 'cheese and pickle'.

Though I enjoyed the laughter that I could generate from such stories, they made me angry, too. The dairy-product-confused psychic could make as many cock-ups as he liked and still play bigger rooms than me. Each preposterous failure would be forgiven; every minor hit was a miraculous success. Fortunately, with even more bizarre and clearly preposterous theories making it into the twenty-first-century mainstream, I barely have time to be utterly appalled by such ghost-whisperers.

Some of James Randi's most damning investigations involved faith healers, a similar scam to psychic mediums, although with the ill being preyed upon, rather than the bereaved being mocked. One of his most famous cases from the late 1970s concerned Peter Popoff, a minister who was stunningly accurate in his healing powers. God would tell him the full names and addresses of the sickly people in his audience, and exactly what was wrong with them. He would then heal them with his touch and watch the donations flood in. Once scrutinized by Randi, it became apparent that this connection to God was via an earpiece that broadcast his wife's specific instructions, based on information that the audience had volunteered on entry. His gift from God was actually a purchase from a high-end electrical shop. The scam was revealed on Johnny Carson's *The Tonight Show*, one of the USA's biggest TV shows. The revelations destroyed the Popoff ministry and he never worked again...

... is what I wish I could write. Sadly, years later the Popoff ministries continue to be active, now under his slogan 'Dream Bigger'. Popoff certainly doesn't want an audience that is too awake – the more docile and close to a dream state, the better. On his donation page he reminds you to donate big, because that is what 2 Corinthians 9:6 tells you to do.[1] God has monetized his miracles. On Popoff's Miracle Spring Water page, you are reminded: 'DO NOT INGEST THE MIRACLE SPRING WATER'. Overly powerful miracles can prove toxic and cause diarrhoea.

When I interviewed Randi about the Popoff investigation forty years on, it was clear that he was still disgusted and disturbed that Popoff could abuse and manipulate people. His voice cracked when he recalled some of the abusive, dismissive and derogatory language that was used to describe the people to whom Popoff's wife was leading him. Not only were they fleecing these people, they were laughing at them as they did so.

Some people may defend faith healing, suggesting that it could have some place in people's physical well-being. If someone is persuaded that their pain can be alleviated through the love of Jesus, it might act as a placebo; but when it comes to cancer, HIV or numerous other diseases, positive thinking alone will not work.

The grand plan is an evil plan, but at least it's a plan

It's good to be a sceptic, in my view, but one criticism of sceptics is that they come across as a bunch of know-it-alls, and that having several sceptics together can lead to sneering and blanket dismissals of other people's more dubious but precious beliefs. I would love to look back on my past exploits and say that I never got an easy laugh on stage via casual dismissiveness and a hearty lip curl (I was the Elvis of the sceptic movement in my youth). But sometimes, the shortest cut to the quickest punchline left ethics at the kerb. Atheists, sceptics and people who have recently discovered that they like modern jazz can all be overly zealous. The trick is to avoid getting a tattoo too soon, as the Thelonious Monk and John Coltrane motifs on my shoulder blades sadly attest to (and I will say no more about the Christopher Hitchens and Richard Dawkins tattoos on my knees).

Sceptical paranormal investigator Hayley Stevens has written about distancing herself from the sceptic movement. She is increasingly driven by an approach 'that doesn't compromise on science, but also doesn't compromise on respect and compassion, either'.[2] It is this reasoned and compassionate approach that seems to underline much of the work of The Merseyside Skeptics. They are the sort of sceptic group that all sceptic groups should aspire to. Obviously that is only an opinion, based on the evidence that I have seen of them so far, which might change as other things come to light. Their reason for being seems to come from a genuine concern for other humans. It is not simply about disproving a theory or belief; it is about working out why people are attracted to thinking that could be damaging to them or others, or where such thinking is just plain wrong.

Revealing that an individual is a charlatan who uses the methods of a con artist to hoodwink their audience only deals with the surface problem; for them, the more important role is to work out how to alert people to the fact that they are being conned, and make the wider world know that it does matter.

The Merseyside Skeptics' co-founder, Michael Marshall, spends a lot of time mixing with flat-Earthers, Moon hoaxers and conspiracy theorists at their numerous conventions. He presents a long-running podcast whose manifesto is to 'examine beliefs from outside of the mainstream, exploring how those beliefs are constructed and what evidence people believe they have to support their case. We believe in approaching subjects with respect and an open mind, engaging with people with different viewpoints in an environment that is polite and good-natured, yet robust and intellectually rigorous.'

Michael wants to understand what has brought people to positions where they consider reason and evidence to be propaganda. It can be a very paranoid landscape that he inhabits. Once you have accepted a flat Earth, then the *Apollo* missions are all lies; and once you have rejected the curvature of the Earth and the Moon landing, you can move on to refuting gravity, the shape of the stars, the structure of the universe – in fact, you can refute anything you want, ranging from vaccines to the existence of The Beatles (Paul McCartney didn't die in 1966, he never existed in the first place. This is the basis of the holographic Mersey Beat conspiracy). All information counter to your position is the product of the Illuminati, an untouchable group who have a bloodline going back centuries, and who are tied to the Knights Templar and to the British royal family. From that point onwards, the world can be whatever dystopian nightmare you want it to be.

Michael talks me through a lecture he saw at a flat-Earth conference, typical of its kind. The speaker simply voiced everything that was in his head – a dirty bomb of misinformation, exploding in so many different directions, with all the disparate information returning to create a single narrative of Illuminati control and paranoia. It sounded like a febrile bedtime story, but what also comes across is a sense of deep unhappiness, anxiety, pathological suspicion and possibly mental illness.*

At one point Michael explained how the speaker described the International Space Station as not being in space. He showed footage of three astronauts larking about in space and doing front-flips. On the ISS, gravity is about 90 per cent of what it is on Earth – the weightlessness observed is due to the space station being in perpetual free fall, or so you or I might believe, but the speaker explained that actually the footage was achieved through the surreptitious use of wires. He focused on one female astronaut's 'wires' and how the footage seemed to show another astronaut pulling her back by her wire. He freeze-framed on 'the wire' and noted that you could not see it, as it had been Photoshopped out, but he knew it had to be there all the same.

What Michael saw, and what I have seen since, was someone being grabbed by their top and pulled in, because the spinning was beginning to get out of control, but obviously Michael's eyes are not correctly attuned for such NASA deviousness. The speaker also pointed to the use of a distraction technique by the third astronaut. Obviously worried that such a moment was going to reveal the wires that everyone was suspended from, the astronaut gets a baseball from his pocket and starts floating it in front of his face. This is the clincher. We are told to consider why anyone would have a baseball in space, especially when it costs

---

* In the first draft of this book I named the speaker, but the more I have looked at their work, the more I am deeply troubled by suspicions of ill health, which is why I decided to keep them anonymous.

so much to transport things up there... if there really *is* an 'up there'. Should he ever get round to reading astronaut Scott Kelly's memoir, the speaker will be shocked to find that he took a gorilla costume into space for a practical joke, while Chris Hadfield famously took his guitar. Although he won't be shocked; the gorilla suit will be just another piece of evidence that space travel is an increasingly flamboyant lie. What is never explained is how this baseball was pre-wired to float in space for exactly such an occasion, or why such footage wouldn't simply be deleted, or how the astronauts are able to float around each other on those wires without getting tangled up. Even if they are not in space, this footage may prove that NASA has discovered wire technology that is far ahead of its time, and which may well revolutionize theatrical productions of *Peter Pan*.

This flat-Earth speech is the opposite of the Occam's razor notion that the best answer is likely to be the one that requires the fewest assumptions. Once you start to follow the threads, you will find yourself almost immediately tangled in another botched perception or presumption, which is why it is frustrating (and frequently pointless) to argue with hardcore conspiracy theorists, because notions of what counts as evidence labour under too many definitions. The conspiracy theorist has an obsessed eye on anything that might confirm their biases, but is totally sightless when it comes to any evidence that may put their position into question. This is not just a habit of the conspiracy theorist, it is a very human reaction. My openness to new ideas will depend on how much they may affect what I already believe.

When observing the errors in reasoning of the conspiracy theorist, it is worth contemplating our own battles to perceive the world as we wish it to be. Which seems about the right time to introduce you to my favourite bagpiping wrestler.

'I've come here to chew bubble gum and kick ass, and I'm all out of gum' – 'Rowdy' Roddy Piper

Although 'The Rock' may be Hollywood's major wrestler-turned-filmstar, the eighties had a few others. You may have a soft spot for 'Hulk Hogan' or 'André the Giant', but I am a big fan of the late bagpiping wrestler 'Rowdy' Roddy Piper. Most would say his finest film is John Carpenter's *They Live* (though I am also very fond of *Hell Comes to Frogtown*, where he is the lead character, Sam Hell, and the town is full of people who look like frogs; it is a pretty nail-on-the-head flick). *They Live* concerns the extraterrestrial reality that lies behind advertising billboards. It is a comic-strip spoof of capitalism, yuppie culture and values promoted by Reaganomics. It also highlights why we may be drawn to extravagant and preposterous narratives bolstered by vast amounts of flimsy evidence.

Understanding why people may furiously deny Moon landings, vaccinations or the birthplace of a president comes back to an idea of control over chaos, even though it is, to me, a rather hideous monster-movie idea of control. In *They Live*, Earth has been infiltrated by disguised skeletal aliens that are turning us into slavish capitalist consumers. Yuppie culture is the work of powers from beyond the stars. It is a comforting viewpoint. The chaos of our lives, our misfortunes, missteps and greed are actually all part of a terrifying alien conspiracy. Your misfortunes are carefully planned and controlled by others. Your faults are not your fault; they are the work of amoral extraterrestrials. Reliable evidence is denied because it is preferable to think that the problems of existence are highly orchestrated and controlled by a greater evil, rather than disorder stemming from the power of nature, terrorism or stupid and greedy politicians.

Which takes us from a wrestler in a fictional alien-infested dystopia to a former footballer who believes he is in a very real alien-controlled hellscape. David Icke's career began as a goalkeeper at Coventry City, but he is now a conspiracy preacher. Icke was a peripheral figure for years, mocked by mainstream entertainment shows, and prolific in his authorship of books with a persecution-complex worldview, such as *Human Race Get off Your Knees: The Lion Sleeps No More*. As conspiracy theory has gone mainstream, so Icke has been elevated and his cantankerous suspicions are à la mode. He has delivered his day-long lectures in arenas, and travels the world spreading the bad news. His spiel is perhaps a bit more sophisticated than that of the flat-Earth speaker, but it still relies on rapid leaps from subject to subject, and a confection of confusions, to convince you.

According to Icke, we are all holograms that are a collective energy of the universe under the misapprehension of our individual consciousnesses, and we are individuals who should rise up. We are ruled by the Illuminati. The controlling group comprises aliens who are humanoid reptilian creatures covered in human flesh, which they can disrobe from in private, and the Moon is a spaceship.

Icke uses disillusion as a prime propellant for his dogma. When your life doesn't make sense, you want an enemy to fight, and a cause to believe in. You reject scientific reality because it has failed to supply you with what you deserve. David Icke was a promising footballer struck down by illness, and then a sports presenter who didn't get promotion to the top job. Somewhere along this line, around the time he became a spokesman for the Green Party, it seems that Icke started to work out what the problem was[3] – and it wasn't him; it was the universe and those around him who controlled it.

In his live lectures he can appear to be encouraging you to question things, but actually he is encouraging you to reject things. It appears that he is offering you a more expansive view, freed from societal shackles, when he is actually limiting what you can accept. Wildly fanciful notions that rest on gossamer-thin propositions become more robust than ideas that have been constructed from years of scrupulous experiment and argument. To dismiss a scientific idea that you do not wish to believe in – whether it is climate change or vaccination – you need no argument stronger than 'They would say that, wouldn't they?' 'They' can be pretty much anyone. 'They' is anyone official, and 'official' simply means someone whose occupation is a geneticist, doctor or virologist. Scientists are bad, maverick scientists are good. The less solid and reputable evidence that you have for your position, the more trustworthy you are. The on/off button of critical thinking is in perpetual motion.

Unfortunately the notion of reptilian humanoids may not empower you; it gives you a story that you can savour and obsess about, but one that leads to a position of 'Why bother?' If the system is controlled by a vast alien conspiracy, there is no way out of it.

Michael Marshall thinks another big draw of these gatherings that reject the mainstream is the sense of community engendered among those who hold such beliefs. Many of the people who take up such stories have felt themselves on the outside, and already possess views of the world that others may consider askew; whilst other people have gone through a personal crisis. When Michael has talked to attendees of flat-Earth conferences, they drop into conversation stories of midlife crises or breakdowns, or talk about how they came across flat-Earth ideas while they were going through a tough time in their lives. For some of them, the crises are now over, but they don't realize that during that tough time their thinking has totally changed track.

Once you are surrounded by people telling you the world is flat, or the Moon that we never landed on is actually a space-ship, then you are experiencing an alternative consensus that is repeatedly confirmed. You are polarized, and you may look back at your former friends who still live on a sphere and think, 'Of course their reality has been continuously confirmed by a society structured on lies – how fortunate that I found a group that can confirm my new, better truth.' There are times when the desire to reject scientific reality is so attractive that it enables you to ignore the problems of your own life. Michael sees that underneath it all, big conspiracies can help us deal with the fact that the world is chaotic and that our lives may well fall victim to many uncontrollable variables.

Such theories were prolific after the First World War, a war when the mechanization of weaponry created previously unimagined devastation. With so much fear and loss, and the return of survivors who had seen such horror and often received terrible injuries, it was hard to accept the world as it was, so there was a need to believe that there must be something more and something different – powers or fantasies that offered an otherworldly solace. The boom in psychic mediums during the years that followed is considered to have been driven by the magnitude of the loss. There were so many things left unsaid that surely there could be ways to communicate with the fallen. Such phenomena as the Cottingley Fairies, photographed by Elsie Wright and Frances Griffiths, may have achieved even greater prominence when they were revealed in 1919 because here was a world in mourning, and desperate for magic. The more powerless you feel, the more confused by how things seem to be, the more attractive the story of our reptilian overlords becomes.

The famous Cottingley Fairies, which captivated so many who were still in shock after the terrible losses of the Great War.

Writing this during the 2020 pandemic, a time of confusion and powerlessness, I have witnessed a year filled with conspiracy theories. It was not long after coronavirus first started to impact on our lives that the belief that the virus had crossed species, because of the poor hygiene conditions in a Chinese wet-food market, was considered an insufficiently good story. The coronavirus pandemic seemed to underline our human powerlessness and lack of control, just as the 1919 Spanish flu brutally demonstrated that, for all the destructive powers of human inventiveness that had been on display in the Great War, nature could still serve up something even more malicious.

For some people, it is unbearable to consider that nature could hold such power, and they believe that only cold, devious humans experimenting in laboratories could achieve such things. This gives us control. All it takes for pandemics to end is for scientists, in their low-lit and devious laboratories, to stop creating diseases. We make ourselves gods – not omnipotent compassionate gods, but those squabbling immoral gods of ancient cultures. If that narrative is not good enough, you move further and deny the existence of a new virus, and turn it into some vicious destruction of society for the benefit of the exalted few. Conspiracy theory is more comforting than chaos theory. Michael Marshall says, 'You end up with a lack of objective meaning because you can't say, this really awful thing happened to me. And it's because of that thing over there. If you can't do that, it's hard to start putting agency back into the world. So I think it's an argument for agency, even if it's a misapplied sense of agency.'

I believe an overlooked part of the initial attraction of conspiracy theories is that having such beliefs gives people something to say: they can hold the floor for a while. Conversationally you are free to hold a highly subjective opinion, whereas opinions on astronomy or the standard model of particle physics have more specific parameters, in terms of reaching a point where you really have to back up your wilder propositions. I have held many opinions based on very scant knowledge and, when questioned on them by people with alternative, fuller sources of knowledge, my natural reaction is to become defensive as opposed to the more suitable reaction, 'Hmm, I should look into this more.' This was Hayley Stevens's eventual conclusion, after spending her youth seemingly being haunted by the ghost of a child-hating old lady.

There's a ghost in my house... or maybe an issue with a chimney

Growing up, Hayley's whole family were convinced that their home was haunted. She was fascinated by ghosts as a child, but also really scared of them. As a teenager, she discovered Internet forums where people talked of their ghostly experiences and she began to become less frightened and more intrigued. This was also a boom time on UK television for spectre-obsessed reality shows like *Most Haunted*.

Hayley believed everything she saw on the show until 2005. The show's star medium was caught faking his spiritual posses- sions. This made Hayley feel 'like a gullible idiot'. Rather than jump to the conclusion that this meant it was all nonsense, she formed her own ghost-hunting team. Two years later she came to the conclusion that 'it was all bullshit', and instead Hayley became intrigued about why people see spirits and why, when people hear a strange noise, they jump to the 'ghost!' conclusion rather than considering that it might be a slavering cellar monster or yet another invisible leprechaun. The knowledge that she had, from having been a believer, has helped her investigation of such claims. She can understand where the believers are coming from. Hayley thinks her first ghost was partly conjured up by the knowledge that a crabby and unfriendly woman had been the house's previous occupant, and that helped whip up her family into a state of believing they were being haunted.

An effective ghost-hunter needs a good knowledge of plumbing, electricity and chimney flues, as well as physics. Hayley's uncle is a plumber and has helped her with many questions about how different pipe and radiator issues can manifest as peculiar noises or draughts. When Hayley's grandmother had an unsettling experience that seemed to suggest a ghost, it was Hayley's knowledge of updraughts and downdraughts that got to the bottom of it. You may have had a similar experience yourself. There were two doors in the lounge: one went into the kitchen and the other into the hallway. One day as her family sat in that room, one of the doors opened, then the other door closed as if somebody had walked through the room. Unsurprisingly, everyone freaked out and wouldn't go back into the lounge until somebody else came into the house and calmed them down.

This was the day when Hayley learnt about air pressure and chimneys. When the wind goes over a chimney, sometimes it can cause the pressure of the flue to drop or change, affecting the air pressure in the room. The likely solution to the ghost-walk was that a wind blew over the chimney and the air pressure in the room dropped, making one door open and the other one close.

Solving the mystery was an exciting moment for Hayley. The existence of ghosts can have testable hypotheses.

When Hayley investigates what may appear to be paranormal activity, she thinks the majority of people are curious as to what is going on, and whilst she might not always convince them, they'll take her answers on board. Those who are hoping for their belief in ghosts to be confirmed may be less accommodating when her answer comes down to a temperamental cistern, rather than an angry spirit insisting on breaking the second law of thermodynamics.

The psychological reasons why people experience ghosts, or even fake evidence of their existence, can be complex. On a TV show I presented a feature on photographs, taken by a twelve-year-old, of what appeared to be an apparition. The investigators that I accompanied soon worked out that a simple manipulation of light and reflection had created a very effective ghostly illusion; but we also became aware of the family situation, which included the recent divorce of the child's mum and dad. The creation of the illusion and the attention it received may well have been part of a reaction to genuine emotional turmoil.

Hayley believes that anyone who investigates paranormal claims should step back and evaluate whether they should investigate them. A lot of the phenomena that are associated with ghosts can reveal underlying medical, psychological or social issues that are causing them. This is one of the reasons why she started to drift away from the sceptic movement. Many people in that movement, when confronted by people with ghostly experiences, will simply respond, 'Ghosts don't exist, therefore it can't be a ghost.' But there are other questions that she believes are important. Why do ghosts persist, at a time when they are so easily refuted and our understanding of reality and methods of investigation are more refined than they have ever been? Why does this person think this is a ghost, when somebody else wouldn't think that? What could be happening in that situation? Are they being tricked? Or is there something else going on with them?

As the work of both Hayley and Michael demonstrates, being a sceptic is not just about disproving events or experiences; it is about understanding what creates them in the first place. A good sceptic is grounded in a fascination with human beings, psychology and the complexity of our minds.

Hayley used to co-present the *Be Reasonable* podcast with Michael Marshall. They would calmly converse with psychics, acupuncturists, astrologers and other people whose claims they would consider to be somewhere between unsubstantiated and preposterous. The agenda of the show, as the title made plain, was to have a reasonable conversation, however unreasonable the hosts might deem their guests to be.

Hayley left the podcast because she found it increasingly difficult to suppress her ridicule when it came to the most ludicrous claims. She got tired of rolling her eyes surreptitiously. Also, as a conscientious sceptic, she found there were so many claims to investigate that she didn't always feel she was informed enough to be interviewing those who were making the claims, so that she could put across good counter-arguments. A further quandary was whether they should be giving these people a free platform. By aiming to avoid being confrontational with them, she would worry that they let the guests get away with too much. Does giving bad ideas sunlight always disinfect them or does that sunlight help them grow?

Hayley still sees ghosts every now and again, but they are not ghosts for long. She was in a graveyard in a Glasgow necropolis, walking around and taking photographs of interesting-looking tombs. She stepped back to get a better photograph and, out of the corner of her eye, saw someone peering creepily around the edge of a tombstone. She freaked out and started to do that walking run that may be adopted by someone who is petrified, but attempting to maintain an ill-fitting composure. Then she stopped and thought, 'You're an idiot. You're an investigator – you investigate this, go back.' She returned and it was simply caused by another tombstone that was slightly askew. It was catching the sun in a way that looked like a head. As her heart continued to pound energetically, she laughed at her fright.

Hayley is now studying for a psychology degree and has found the untrustworthiness of eyewitness testimony, and the wariness that we should have when dealing with what our senses tell us, particularly illuminating and useful for her own studies. We are able to build up complex narratives on very little actual information. It is important to remember that all narrators are unreliable – and that includes you.

'Science… is forever whispering in our ears, "Remember, you're very new at this. You might be mistaken. You have been wrong before"'[4]

By this point you might still be wondering why all this matters, thinking that we should let people believe in nonsense if it makes them happy – although from my experience of online conversations, it rarely seems to make them happy. Frequently my experience online has been to witness an aggressive righteousness. The doubt that comes from well-aimed scepticism really becomes important when it comes to our lives, our families and – on climate change – our descendants. What might at first appear to be the killjoy assassination of gods, ghosts and monsters becomes far more important when we see how it can be exercised for the vital decisions that we may well have to make over medical progress concerning the future of humankind.

One of the Merseyside Skeptics' lengthy investigations has been into alternative cancer therapies and 'cures'. The danger of street-corner purveyors of dreamscape cures and tree-root memory potions comes brutally into focus in the case of Sean Walsh. Sean was a young musician in possession of a wonderful spirit of chutzpah and joie de vivre. Diagnosed with Hodgkin lymphoma as a teenager, he underwent six months of chemotherapy, which defeated the cancer for a while. In his early twenties it returned. Having experienced the exhausting regime of chemotherapy once, it is not surprising that he was attracted to unofficial alternatives that promised recovery while causing little discomfort.

The Internet is filled with false hope. Sean was determined to do it his way, and he rejected what the National Health Service advised and found himself at the mercy of all manner of quackery – some benign and some that was, ultimately, deadly. One of his guides was someone who had cured herself of breast cancer, or at least believed she had. According to research collated by The Merseyside Skeptics, there is no evidence that she actually ever had cancer, as she was never diagnosed.

Michael Marshall sees people becoming 'radicalized' by those peddling alternative cancer treatments. It is a conspiracy-theory mindset that can lead to all officially tested healthcare being seen as part of a grand scam. Sean was a charismatic individual, and his public devotion to alternative therapy made him something of a celebrity in the alternative community. One clinic was still using him to publicize their integrity and effectiveness long after he had died. When Sean was diagnosed, the doctors had given him a fifty/fifty chance of living a long life, if he received proper treatment.

Among the alternative treatments he had was thermal imaging. He believed, from the thermal images, that his tumour was shrinking, but thermal imaging is not an effective way of monitoring tumours. By the time there was no denying the progress of his cancer, it was too late for conventional medicine to make much impact at all. Sean's family and girlfriend are now campaigning against the alternative-therapy and treatment businesses that, at the time of writing, are still trading. His partner says that she feels she lost her boyfriend to cancer conspiracy theories.

Lorna Halliday's mother had breast cancer. She was highly suspicious of conventional medicine and so she kept her diagnosis secret from her children, believing they would try and convince her to take that route. She was given a good chance of survival if she had a mastectomy, plus chemotherapy and radiotherapy. Instead she had mistletoe injections and was given a caustic ointment of black salve, which created a scab on her breast. When the scab came away, she was told that was the cancer coming away too. Instead her cancer became an ulcerating tumour and she died of septicaemia.

On the *Skeptics with a K* podcast Michael said, 'This is a story of radicalization and of how a constant dripfeed of exposure to pseudoscientific claims can lead people to extreme ideologies.

That radicalization is even more effective when it comes to vulnerable patients.'

Theoretical physicist Brian Greene believes that to uncover any deep truth, you have to be comfortable with uncertainty and doubt. You need to build a solid foundation of uncertainty. For him, this is a vital part of the educational curriculum: in science classrooms, in every classroom. He believes we must stress to students that there is beauty in doubt and there is a beauty to uncertainty. He describes it as 'the opportunity for us to make headway, to make progress, to incrementally understand the world a little more deeply, to read a little bit of the doubt that came to us from an earlier generation of thinkers. That's an exciting journey. That's why we get up in the morning and want to go to work, because of the doubt and the uncertainty that we can wrestle with. I think if you get students at a young age, they can really make that part of their worldview – they're part of how they are existing in reality, and that would be a vital step forward.'

We have to find ways of persuading people of the attractiveness of scientific doubt and the destructiveness of certainty. Looking at the length of David Icke's career as a prophet, it is clear that mockery is not a solution, it is merely a release for frustration.

Ann Druyan, often with her husband Carl Sagan, created some of the most passionate and accessible works of science communication in the late twentieth century. Her work includes co-writing some of the powerful essays in *The Demon-Haunted World –Science as a Candle in the Dark*. She stresses the need for humility if we are to confront our place in the universe, and reach out and connect without ridicule. We need to accept that we cannot know everything, and we need to be open to listening to reliable sources that may well shake our confidence in our presumptions. We should be wary of any self-contained belief system that is immune to logic and immune to evidence. As the conspiracy mindset thrives in the mainstream, Anne concludes, 'We are so clever. And so stupid. And it's really quite amazing. And I think, at this moment, we are as clever and as stupid as we've ever been.'

Part of the cure is ensuring the accessibility of science, and ensuring that comprehensible accounts are available. Another part of it is encouraging people to read well, to find time to walk beyond their comfort zone. There is frustration when you realize that some ideas are hard to understand. I have grown to accept that the best I will ever manage is a very shallow understanding of the laws of nature and of the physics beyond it. I do not count that as a surrender – I'll keep working at it, but I face the reality that I can have fun imagining and playing with scientific ideas, and I will be able to colour in some of the universe in my head, although a Nobel Prize is not looming.

I think that some of the pseudoscience, and the anger that inhabits it, comes from the refusal to acknowledge that some people may be smarter than you – so rather than accept that or concentrate on understanding equations, you declare that those who have the gall to appear to know more than you have made it all up. Then you don't need to bother learning their ways; you can simply get on with tinkering with the cold-fusion machine you have in your barn, or the cure for all the plagues that you are making out of parsley and Sprite.

Some call them truth-seekers, others call them Covidiots

2020 was a year when the importance of strenuously exercised scepticism became very apparent. It was a year filled with doubt, not all of it useful doubt and much of it doubt that rejected anything that might counter a desired narrative. The Covid-19 pandemic was a sweaty incubator for conspiracy theories, and the playing out of paranoias has been centred around coronavirus, feeding the fevered imaginations of the conspiracy-minded with copious shovelfuls of hot coals.

On the efficacy of masks, the effectiveness of lockdowns, even the existence of the virus itself, there have been many with megaphones projecting their truth. I can see why this would be such a rich time for aggressive paranoia. If you are not affected by Covid, if you have not seen or experienced it and you have been informed by those who – for the purposes of cash or ego, or both – are typing out doubt, I can see why you might like to believe that even if it is happening, it really can't require all those draconian measures. At the time of writing there have been more than 128,000 Covid-related deaths in the UK, as well as many people suffering from the effects of long Covid, and yet despite this, every now and again I think, 'But is it really so bad?' Then I remind myself of the nurses and doctors I know who have told me how things are on the front line; and of the virologists and immunologists that I am fortunate to have access to, and what they have told me. I ask myself: would this number of experts, in so many areas, all be so wrong, and the shock-jock DJs and media pundits, with no background or education in medicine and disease, who have so often been wrong in the past, turn out to be right? That someone with access to such information can still have moments of active doubt during the pandemic makes me realize how easy it might be to fall down a rabbit-hole of disbelief, especially when we are perpetually submerged in the media of constant anecdote. It is at a time like this that I understand that scepticism often requires a battle with minds that can easily fall into lazy supposition and paranoia.

My neighbour believes that Covid-19 was made in a Chinese laboratory (something I doubt but perhaps by the time you read this, there will be irrefutable proof). People I have worked with believe that Bill Gates has placed time-travelling nanorobots in the vaccines, to control their minds. Others I know suggest that the wearing of a mask is the greatest imaginable attack on their civil liberties, which I find odd, as people in showbiz usually like dressing up.

Rupert Beale is a doctor who runs a cell biology lab and was one of the first experts in the field of virology and immunology that I spoke to during the pandemic. In his first column about the situation for the *London Review of Books* in March 2020 he wrote:

1. This is not business as usual. This will be different from what anyone living has ever experienced. The closest comparator is 1918 influenza.

2. Early social distancing is the best weapon we have to combat Covid-19.

3. Humanity will get through this fine, but be prepared for major changes in how we function and behave as a society until either we're through the pandemic or we have mass immunization available.

When I talked to him almost a year on, one of the first comments he made was about how successful the vaccine research and development have been, in less than a year. Not only was there a successful vaccine but, despite the more persistent vaccine-deniers, Rupert had seen a very good uptake – even higher than he might have hoped for.

From a point of pure speculation, he explained that this might tell us something about the psychology of these situations: 'people are quite happy to refuse vaccines when there's no obvious benefits. And very often it's not a question, let's say, for the flu vaccine. There's loads of people over the age of sixty-five, or who are healthcare workers in contact with lots of people, who just don't bother every year. It's quite hard for them to see the consequences of that. And I guess, with the lockdowns that we've had, and the misery it's brought, it's sort of obvious to everybody what the consequences are of allowing an infectious disease like this to run unchecked.'

Of those who are eligible for the vaccine so far, about 5 per cent have decided against having it. Rupert wonders if 5 per cent is the magic number of hardline conspiracy theorists. The problem then becomes trying to ensure that good information is disseminated to make sure that 5 per cent don't infect more people with their unfounded suspicions. I find it very depressing when I see so many people rejecting evidence-based action, especially when it is about protecting ourselves from something as potentially life-threatening as a pandemic, but Rupert believes that the problem feels worse due to the shrill persistence of a few people on social media and other forms of media.

Rupert did see vaccine hesitancy and uncertainty – and even denial of Covid's existence – from people who had lost a family member or seen someone die in the hospital bed next to them, but over time most people were persuaded, especially as it became increasingly apparent that serious side-effects from the vaccine were highly unlikely.

The anti-vaccination movement in general has gained much ground over the last twenty years, based on considerable misinformation and disinformation. At times the lies about the MMR vaccine for measles, mumps and rubella, and the downturn in the number of people having the jab, saw the UK coming close to losing the ability to control measles. Rupert hopes that the success of the vaccines against Covid-19 may go some way to restoring the public's confidence in vaccines. He believes that 'We'll see a fantastic change; people will be very much more pro-vaccine. I think that will make a big, big difference.'

The important thing is to get good information out to people very early on, and he does feel there was a failure of scientific advice in the early stages of the UK's Covid pandemic response. He has also seen that some people who have been given the title of 'expert' by the media – although Covid might not really be their area of expertise – have continued to be allowed to publicize their beliefs. In my naivety, I remain perplexed when people that I presume to be wise find it so difficult to retract or retreat when the evidence is just not there (I know, I am *so* naive).

Rupert believes this is because 'They say something, which turns out to be massively against the public-health interests of a population. And you carry on repeating that, so that for you to change your mind requires you to accept that you have probably caused people to die. It's actually a very difficult thing to realize that you yourself have not only made a mistake, but have done

50

something which is potentially really seriously detrimental to a massive public-health effort. And I think that's a very difficult thing for anyone to accept about themselves.'

Dr Emma Hodcroft studies and tracks viruses and is co-developer of Nextstrain, a real-time tracking system for pathogen evolution. She has found that the pandemic has required rapid education in how best to talk to the media effectively. She believes that the initial science communication may have fallen flat. In the early days of the pandemic there was no real evidence that masks were effective at slowing the spread and, with a worldwide shortage of protective equipment, it would have been ill-advised to tell everyone to rush out to buy them, and would possibly make it even more difficult for medical workers to obtain them. What Emma sees clearly now is that scientists may be very aware that when the evidence changes, the needs and advice may change too, but people can remain stuck with the first story they heard. This means that new and changing information may need to be repeated over and over again.

She has spent some time thinking about this problem so that it can be handled better in the future. She believes that scientists could probably have been more open from the beginning about why masks weren't being recommended, 'and definitely been more open that this is something we're investigating, and that the science might change on this'. Emma thinks this communication improved as the pandemic continued. Now, in interviews, she will repeatedly stress that research is ongoing and that the advice might alter. The aim must always be that advice is the best advice possible based on the evidence available, but that evidence – especially in something as large as a global pandemic, where so many scientists are focused on research – may change rapidly.

One of the most important messages that she hopes has become clearer to the public is that science is an ongoing process, and that there is no harm in repeating this. Stay tuned: you never know when new information is just around the corner. The repetition of simple concepts is really important for communication, as is clarity. The best counter to misinformation is good, understandable information.

Emma thinks that much of the reason why conspiracy theories are attractive is because they are understandable. They appear to make sense, and you don't have to have a degree in anything to comprehend most conspiracy theories. Reality is often more complex. We also have a climate in which people are offended by the notion that other people – actually experts in the field – may know more than them. It means that it can be hard to communicate accurately in a way that is comprehensible and is not perceived as patronizing. She explains, 'I think we could be having better outcomes in how we communicate our research... But it will take convincing scientists that it's worth it, and giving them the resources that they can take the time to do it and that they know what they're doing.'

The promotion of effective doubt, and the successful communication of information that could save lives, requires responsibility and humility. This is not simply about the humility of scientists, but about the humility of all of us. We need to accept that we are not experts on everything, and that there is no shame in knowing less about a virus than a virologist, and less about pathogen evolution than someone who spends a lot of time studying it. It is also about responsibility: if you make extraordinary claims about a disease or the height of a cliff face, then you have to take responsibility if people become infected or if they crash onto the rocks below because of your claim. Personally, I think the news media should be a little more wary of who they give the megaphone to in such life-and-death situations, and that the victory of Twitter trending may be costly for some.

The unexpected polar-bear population expert

As a sceptic, it is important to remember to be sceptical of sceptics too, though there are some sceptics that I am more sceptical of than others. One such group would be people who claim to be climate-change sceptics when often they are really ideologically opposed to any evidence that conflicts with their certainties.

I have been appearing at events with Helen Czerski for many years. She is one of those scientists who seems to be able to adapt to any question, whatever the discipline. Her one weak spot is films. I have found the only way of getting the upper hand, when she is expertly dissecting sunspots or the uses of whale excrement, is to bring up *RoboCop** or *The Thomas Crown Affair*. We presented a few podcasts together about Covid-19, and the obfuscation that we saw there brought us back to another area where the need for urgent action is slowed down by misinformed resistance.

Helen has been on numerous Arctic expeditions and is someone I often turn to when trying to understand climate change – another area where many people reject much of the science and

* Obviously the original *RoboCop*, not the remake, which managed to surgically remove everything that made the original such a masterpiece.

where, indeed, sometimes they are paid to do so. Although there are climate-denial zealots, there are also many people who are simply swept along with a movement. The 'climate-change scep-tics' have often been given a lot of room to play in the media, and fossil-fuel companies have funded individuals and organizations to cast doubt on the scientific consensus.[5] There is a huge amount of uncertainty about what to think on climate change for many, because one of the aims of the denier movement is to sow confu-sion and uncertainty. With an issue as big and as potentially cata-strophic as rapid climate change, why are so many of us resistant to so much evidence?

Helen believes it is important for science to be sympathetic to the idea that most people don't want to hear about climate change. It's not because they are terrible people, but because it's climate change, it can sound like it is all their fault. It can feel as if we're being told off and, as we all know, when you are being told off, you switch off or wave your arms around and say, 'This is *so* unfair.' It can feel as if the scientists are going to come along and take all our fun away; as if they are going to stop us going to places and doing things, and eating what we want, and they are going to wag their fingers at us.

Sadly, I am often one to overreact at the slightest hint of admonishment and become rather brattish in declaring, 'Oh, so you are saying this is all *my* fault', so I understand this feeling. Whether it is failing to remember to buy hot-cross buns or con-tributing to the end of human civilization, some of us can get uppity. Helen compares it to the reaction that vegans get when they tell meat-eaters they are vegan. The meat-eaters – rather than hear, 'No, thank you, I won't eat the lamb... No, it's not that it doesn't look delicious. Sorry, I'm a vegan' – often hear, 'You slaughtering scum, with the blood of sheep on your hands and lips. You disgust me!'

Helen has seen very few scientists take such a high-handed and judgemental stance, but the idea that they might has stuck, so many of us are offensively defensive about climate change and our role in it, defending our way of life. She worries that scientists haven't been very adept at simply being human about it and understanding that people might have good reasons for wanting things to stay as just they are. And now the further problem is that it has become tribal and, once you are in a tribe, it doesn't matter what the other tribe says: they *are* the other tribe, so they are wrong.

There are still many institutions that view economics as mattering more than the climate. The idea that much of the Earth will become uninhabitable for humans seems too fanciful, too far away to worry about. Helen doesn't think scientists have been set up to deal with this tribal thing; they still think the rational information they provide will persuade people. She believes there is a need for the emotional story about climate change to be at the forefront of the argument. She thinks the solution comes down to listening to people: 'Let's listen to the people who think they're in the other tribe and find out that they're not evil. They're not setting out to destroy the world.' For many scientists, there may be a feeling that science is getting the facts and what is done with those facts is nothing to do with them, but this is changing. There is a new need to translate the information, especially when other institutions will distort the narrative to protect their finances.

The problem now is that there need to be faster ways to have these discussions, because action needs to be taken. But action is hard; we can spend a whole day reading about climate change and the need to conserve energy, and then feel too tired from all that education to go and turn off the bathroom light.

It is not the tragedy of science, it is the tragedy of humankind

The importance of critical thinking, of doubt and of searching deeply within ourselves to work out why we believe what we believe is brutally put into context by a sequence from one of the greatest TV science series, *The Ascent of Man*. A sixty-four-year-old mathematician in a suit and tie stands in a field on a grey day and walks purposefully into a shallow muddy pond. He squats down and plunges his hand into the murk and weeds, then drags up a handful. This is Jacob Bronowski.

This particular episode began with a contemplation of how an artist captures a face, moved on to our understanding of light waves and then on to the discovery of the theory of quantum indeterminacy – which is something I would sum up if I were a cleverer human but, like much in the theories of quantum physics, it is pretty difficult to summarize it breezily without jarring your mind and going into a fug of incomprehension. Bronowski's summary was that if we know where a particle is, then we don't know its speed or direction; and if we know its speed or direction, then we don't know where it is. The more you know one property, the less you can know the others.

The theory fuels my favourite physics joke. Quantum physicist Werner Heisenberg is speeding down the highway. A traffic policeman pulls him over. When Heisenberg rolls down his window, the cop asks sternly, 'Do you know how fast you were travelling?' Heisenberg replies, 'No, but I know exactly where I am.'

What Bronowski takes from this theory is that our knowledge is limited, and from this quantum lesson it is important to remember this limitation. At the same time as the quantum pioneers were revealing the limitation of our knowledge, Europe was descending into one of its most terrible periods of brutal intolerance. Truly great minds of the twentieth century, minds that would change our understanding of the world – Freud, Einstein, Schrödinger – were considered to be subhuman. The realization that all knowledge was limited was, tragically, in direct conflict with a regime that would murder and destroy, justified in their actions by their own twisted dogma.

The pond that Bronowski walked into is in Auschwitz. It was here that members of his family and his friends were murdered. 'It's said that science will dehumanize people and turn them into numbers. That's false, tragically false. Look for yourself. This is the concentration camp and crematorium at Auschwitz. *This* is where people were turned into numbers. Into this pond were flushed the ashes of some four million people. And that was not done by gas. It was done by arrogance, it was done by dogma, it was done by ignorance.'

Twenty-eight years after this mass extermination, Bronowski warned us that the belief in absolute knowledge can only ultimately be destructive. Never has the importance of doubt been so eloquently and tragically expressed. Well-used scepticism – scepticism that probably engages with what it means to be human, that combines scientific knowledge and compassion – far from creating a cynical world, gives us the chance to increase our empathy, to challenge ourselves, to move forward.

I can live with doubt and uncertainty and not knowing. I think it is much more interesting to live not knowing than to have answers that might be wrong. If we will only allow that, as we progress, we remain unsure, we will leave opportunities for alternatives. We will not become enthusiastic for the fact, the knowledge, the absolute truth of the day, but remain always uncertain... In order to make progress, one must leave the door to the unknown ajar.

RICHARD FEYNMAN

CHAPTER 2

Is God on Holiday? – Are There Still Enough Gaps for a God?

> One golfer a year is hit by lightning. This may be the only
> evidence we have of God's existence.
>
> Steve Aylett

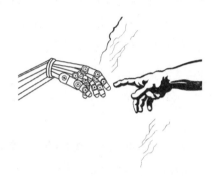

The problem with too much interrogation of the universe is that it can undermine the standing of a deity, and if it is a deity you feel close to, this could be a thorny problem. But does science have to extinguish your gods?

Where are the gods in the scientific view of the cosmos? For many, there comes a point where a universe of provable physical laws defined by equations extinguishes the need for a greater power shaped like a god. One of the most antagonistic feuds between science and religion takes place in the Garden of Eden. It is not cosmology that butts heads with religion, after all, as it seems the universe has a beginning and such a story can fit with the scriptures, so long as you ignore the timeframe. You need more than 7,000 years to get from nothing to this much of something. Biology is a more problematic concern for the dogmatic. Removal of the 'Made by God' logo on the sole of our foot is a step too far. God as the toymaker of all that lives on Earth is an idea that the more fundamentally religious demand. They demand to have been smelted in the blast furnace of heaven.

The 2009 film *Creation* tells the story of Charles Darwin coming to terms with the loss of his beloved daughter Annie, whilst developing his theory of natural selection. In one scene a pugilistic Thomas Huxley tells a nervous Darwin, 'You've killed God, sir.'* Darwin looks perturbed. He knew that the power of his theory of natural selection would do great damage to the concept of God. He wrote to his friend Joseph Hooker and admitted that he feared his theory was 'as if one were confessing to a murder'.

In later life, Darwin no longer attended church. He would walk his family to the door, but would not go in himself. Though his scientific theory may have played its part, there was also the terrible effect on his religious beliefs of losing three children. Emma, his wife, worried deeply in a letter to Charles about his loss of faith, fearing that it might divide them after death.

---

* Surprisingly, *Creation* was distributed by a company founded by a member of the Catholic Traditionalist Movement, Mel Gibson. The screenplay is based on the Randal Keynes book, *Annie's Box*, about Darwin's work and the box that he kept, containing memories of his late daughter. I have a few quibbles with the film – most of all, the portrayal of Emma Darwin. Read the book to see some of the heart-wrenchingly beautiful correspondence between Charles and Emma. I thought she was a bit too grumpy in the film. It is a love affair that deserves a film of its own.

Charles replied to the letter, writing, 'When I've died please know that many times I've kissed and cried over this.'[1]

Does science kill God?

The brusque younger me would have imagined it is inevitable that the evidence-based view of the universe and a belief in God cannot exist in tandem. I have performed a few stand-up routines about Creationism, fundamentalists and the Bible over the years. I've told the tale of a toxic preacher sitting proudly on a high horse of relentless homophobia who, in mid-flow, was interrupted by a seagull. The gull had clearly eaten something from the dustbins that had violently disagreed with it, and the preacher was soon spattered like Jackson Pollock dispensing ice cream. Some would say that the volume of avian diarrhoea that landed upon him was clearly a sign from God. We may think of the Dove of Peace when we think of the ornithology of the Bible, but maybe the Gull of Gastroenteritis came from one of the gospels that disappeared during the early days of biblical edits.

In the last few years, though, I have become increasingly fascinated by how complex some people's ideas of God are, and their relationships with Him or Her or Them – or, as it frequently seems, an idea of God that becomes increasingly vaporous. Such deities cannot merely be evidenced away, even by the brightest people. Equally, my own lack of faith in God did not come from intense reading of evolutionary theory or cosmology; the loss came first, the scientific reading later. Just as I did not require science to remove God from my rituals, so there are those working within science who do not find that it needs to extinguish their god.

Nevertheless, I still find myself surprised when I meet scientists who have a traditional faith. I should really have got used to it by now. Here are people able to comprehend ideas way beyond my capability, but among all those formulae and deeply considered notions about the curvature of spacetime or the development of eukaryotic cells, there is room to fit a god, too.

Fundamentalists are easy to understand and easy to dismiss, though this may well not stop their aggression and, sometimes, their violence. Their angry, woman-demeaning, gay-hating god of hypocrisies is a concrete cliché, an escape from having to reason with their bigotries. As the author and activist James Baldwin wrote, 'I imagine one of the reasons people cling to their hates so stubbornly is because they sense, once hate is gone, they will be forced to deal with pain.' When God doesn't provide for such specific needs – when a god is elusive and inscrutable, and far from physical or visible as a bearded bigot on a throne – it starts to get trickier.

Science doesn't disprove God, it just gives God a lot less to do. This is the god of gaps, and whilst his creativity and imagination are no longer required for all explanations, some smaller holes are left to be plugged.

I was brought up in a churchgoing family, but there was no fundamentalism. The level of religious obedience meant that my older sisters were scared to take me to see *Monty Python's Life of Brian*, and there was no way we would have been allowed to sleep in the same room as our partners even a good ten years into a relationship. But most of the time my family's religion remained pale and in the background (apart from the monthly sacrifices of the fattest sheep, and occasional bouts of speaking in tongues after repetitive acts of flagellation with the blessed whips, obviously).

My memory of a personal God is blurred. I wish I could picture what I saw when I prayed to him, but perhaps I didn't see anything. I was unthinking in my religion. God was a received notion. It was not Darwin or Dawkins who destroyed my god – God just faded out of sight. I did not argue God out of my existence; God simply wasn't there. I have no God-shaped hole, but simply a spare hour to fill any way I like on Sundays.

I am probably an atheist; maybe I could be called an agnostic. I wouldn't even mind pantheist – it's not really of any concern to me. It is occasionally frustrating. On bad days, it might be nice to look up at the sky and scream about being forsaken, but perhaps it is my own fault that I have hammered my thumb with a mallet, or perhaps it's the fault of gravity, or I can blame some notion of the chaos of the universe. Whatever, I must aim to own my own incompetence, although I did blame Zeus once when I was attacked by a swan on the Grand Union Canal.

When we recorded the 100th episode of *The Infinite Monkey Cage* we introduced the idea of a 'clerical corner'. The show was an overview of what we had learnt about the universe in the eight years since it began. In between scientists extemporizing about particle acceleration and light-swallowing singularities, we sought the opinion of two real-life clergymen: Victor Stock, the former Dean of Guildford Cathedral; and the Reverend Richard Coles, former keyboardist with the Communards and an occasional podium dancer in his hedonistic youth, before turning to tapping his foot in a pulpit. They both represent the sort of clergymen you would hope to populate the Anglican Church: brimful of curiosity, poetry and doubt. Either could have been comfortably portrayed onscreen by Alastair Sim.

The astrophysicist Neil deGrasse Tyson was also on the show. Coming from the USA, he is used to the more aggressive attitude of the monetized, evangelical brimstone preachers of the American Midwest, the sort that spew Stygian greed and promise Hadean agony, before being found in a motel room, filled to the gills with poppers and placing the collection-plate cash in the G-strings of a trio of rent boys (they get away with it, too, with that cast-iron 'the Devil made me do it' alibi). The first time we introduced 'clerical corner', Neil jovially sniped, 'Do they have a science corner in all of their churches?'

The avuncular Victor, who has immaculate delivery with a perfect impish glint of camp, replied, 'Well, you see in Westminster Abbey, where we have Stephen Hawking, Isaac Newton and Charles Darwin, one of the things we aimed to do is...' And he continued to praise evidence-based curiosity in the perfect pulpit manner, creating a shaded area on the Venn diagram where faith and science overlap. All further sparring was made redundant. It was a happy reminder that your position in the pew or the laboratory doesn't instantly mean you must be at loggerheads. Dogma is the enemy of evidence-based thinking, but religion doesn't have to be.

'God created physics and then went on holiday'

Carlos Frenk is one of my favourite cosmologists to talk to about God (and also about Ingmar Bergman and René Magritte). He is based at Durham University and has spoken at the cathedral with the then-Bishop, David Jenkins. Jenkins was the *bête-noire* bishop for orthodox Christians, whose backs went up when he went on television to say that he didn't believe in the literal truth of the virgin birth. There was a campaign at York Minster against him being consecrated as Bishop of Durham. Three days after his service of investiture, the Rose Window of York Minster was struck by lightning. Some declared it to be the righteous hissy fit of a stroppy God. (Before you anger the Almighty, it is always a good idea to make sure you have the number of a good glazier.) Jenkins thought that no one should believe in the kind of God who would care more about the opinions of academic theologians than about the suffering of children in Auschwitz. Why should lightning strike York Minster when God conspicuously failed to stop the Holocaust?[2]

The theme of the Durham Cathedral cosmology event was the creation of the universe. There were speeches by an artist, a biologist, a geologist, by Carlos and the Bishop of Durham. Carlos saw the Reverend Jenkins as an intellectual, and intellectuals are frequently frowned on by the British. It seems that we often prefer confident ignorance.

Carlos was the penultimate speaker at the event. The origin of life, the formation of the Earth and the meaning of creativity had all been covered, so he explained his ideas about how the universe came into existence. When he sat down, the Reverend Jenkins took to the podium with great fanfare. He stood before the audience in full regalia and said... nothing. He looked around the room and at the audience and continued to say nothing. It was beginning to become embarrassing. The audience was expectant. Silence is a hard thing to hold. Think of all those conversations you have had with minicab drivers when you would really have preferred silence. Even the most resolute monk fights the urge to say, 'It's been a funny old day for weather' when the abbot walks by.

After two or three minutes the bishop said, 'Silence is important. I came to this debate to learn, not to teach.' With that, he walked off. He had the biggest impact of everyone that evening, putting everything else into context. Carlos told me, 'Months later – years later – people will remember the silence. He encapsulated everything; all my beliefs in God were encapsulated in that moment, in three minutes of silence.'

I had met Carlos and sat on panels with him on numerous occasions, but one day I heard him explaining to a radio host that he 'does not allow God into the laboratory'. This took me by surprise. What sort of god did he mean? Was he being literal or playful? Was this the god that was talked about by Einstein and Hawking – the one who doesn't play dice with the universe – or a god whose mind cosmology will allow us to know? As far as I could see, this was not God as metaphor for knowledge or physical laws, this was God as a god. Apparently God prefers not to be a metaphor, but when the big guns of physics mention him, he'll take any publicity he can get.

Carlos was brought up half-Jewish and half-Catholic, in a German family living in Mexico. Being brought up in both these religions, he carries the weight of both worlds with him: two burdens of guilt for the price of one. Between the two possible deities, it is the Jewish God he feels close to. He sees the Jewish God as 'a presence that you interact with on a daily basis'.

As well as having a god to walk with, he is watched over by Albert Einstein. A large bust of the physicist sits imposingly on a high shelf in his home, looking down on Carlos with suspicion. The study of cosmology is everyday nutrition to Carlos, helping us to realize how small and how irrelevant we are. I find his ability to exclude God from his experimental and theoretical thinking intriguing. How are you able to interrogate the laws of the universe, but keep Him out of it? Then, when the laboratory lights are turned off, He's allowed to accompany you on the walk home? It must be a tricky business telling God to wait outside, like a child left outside the pub with a packet of crisps.

Carlos delights in the universality of the laws of physics and how they seem to apply at all places and all times. He wonders if you could say that this delight has nothing to do with the universe, but that it's to do with your brain – that it is a construct trying to make sense of what you perceive, though the universe couldn't care less? The most remarkable thing he believes is that he can predict events before they happen, predict the future in a way that would make Nostradamus turn purplish with envy. The affairs of humans may still be unpredictable, but the affairs of the stars and the galaxies can be predicted, not through a crystal ball, but through the prisms of telescopes and the scientific knowledge they give us. For Carlos, this is the most incredible thing about the universe. This is where his god comes in. He believes that God's act of creation was to create physics, and after that He went on holiday. God's holiday also helps us understand the problem of pain and suffering, something that always seems to me to be the stumbling and coughing point during conversations about omnipotence and omniscience.

Our battle with this contradiction is illustrated by the parable of the trial of God at Auschwitz, which, according to Nobel Laureate Elie Wiesel, was more than a parable. It happened. This was the night when those suffering the vile agonies and hideous indignities hurled at them by the Nazis, because of their faith and their culture, demanded to know why their God would simply stand by. Wiesel told an audience at a Holocaust Educational Trust appeal dinner in London, 'I was there when God was put on trial.' When this was questioned by rabbis and academics, Wiesel replied, 'Why should they know what happened? I was the only one there. It happened at night; there were just three people. At the end of the trial they used the word *chayav*, rather than "guilty". It means "He owes us something." Then we went to pray.' As Teyve says in the musical *Fiddler on the Roof*, 'I know we are your chosen people, but once in a while, can't you choose someone else?'

Nature is full of pain and suffering, of offspring who must be destined to die young, of the mechanisms of survival for one species inflicting agony on another species. If God was the toy-maker of all this life, then he had days at the workbench when He was full of malice.

David Attenborough has discussed his reaction to the question 'Why are we here?' People will sometimes say to him, 'Why don't you admit that the hummingbird, the butterfly, the bird of paradise are proof of the wonderful things produced by Creation?' Beauty in the universe is the bargaining chip for the existence of God. Attenborough replies, 'You've also got to think of a little boy sitting on a river bank in West Africa that's got a little worm, a living organism, in his eye and boring through the eyeball and slowly turning him blind. The Creator God that you believe in, presumably, also made that little worm. Now I personally find that difficult to accommodate...'

Carlos worries about this too. How can there be a God, when people are exploited, starving and in physical pain? How can God allow someone to live a miserable existence? If God created the rules and then the events happened, and one by-product of these rules is our existence and our conscious knowledge of this God, then surely this God would have been aware of the implications for 'his' species?

When he was growing up in Mexico, Carlos found it hard to make sense of the Catholic teachings while there were children dying of hunger. On the Jewish side, how could he make sense of the tremendous injustice? Is the god who departed on holiday, after setting everything up, something of a cop-out? He thinks he can make intellectual sense of it all, but worries about the objection, 'What if that is not true, and physics only exists in my head and it doesn't exist out there? Then it gets a lot more complex.' He also worries that a god that writes a physics textbook and then goes on holiday is not a very interesting being. What has he been up to all these years?

I drag up the usual argument from my Bertrand Russell pocketbook of atheist arguments: how can it satisfy us to say that the existence of the complexity of the universe came about due to the actions of something even more complex? Carlos suggests that there is no reason for God to be complex. But if God is not complex, then he is not the god that the majority of religious people think about when they think about God. If God is not complex, who are we having a conversation with when we are kneeling in the mosque, synagogue or chapel? You may pray for your mother's sickness to be cured, but would that require God to break his own laws of physics? Carlos fears that from there, chaos would ensue.

Despite all this, he sees his belief in a greater power as an intellectual necessity, and says that if he lacked this dimension he would be lacking something very important. Carlos explains that it is intimately related to his studying of nature – that it doesn't make sense unless he has this framework. When he talks to God, it is a one-sided dialogue. He's never heard 'the voice' or seen the light. He considers himself to have an amicable relationship with whatever created the universe.

For me, God is the voices in my head, if he is anything, and that means I keep a lot of gods hanging around in my skull, and a lot of them come up with terrible ideas. As the great comic-book writer Alan Moore wrote in *From Hell*, 'The one place gods inarguably exist is in our minds where they are real beyond refute, in all their grandeur and monstrosity.'

Carlos believes you should turn your back on irrationality in daily affairs, but you shouldn't turn your back on irrationality in your private life. Irrationality is a legitimate part of humanity. Private irrationalities come with our imagination. Sadly, many people seem to want their private irrationalities to be public, and they want you to obey their irrationalities, too.

I am attracted to this messiness. Here is someone who has a depth of scientific understanding about the universe that I will never have, a grasp of its complexity that evades me, a way to comprehend an equation that would send me reeling, but he needs his irrationality too. Those tears that come to

Carlos's eyes when he thinks of the discovery of cosmic microwave background radiation are part of that story to him. 'The human mind was capable of predicting this phenomenon that was then verified empirically. Why is the mind capable of taking everyday experience and extrapolating? This encapsulates what science is about, and what God is about. I don't see how that can happen unless there is an agent… an order that transcends everyday life.'

'Touching the hem of transcendence, in a moment of eternity'

I first met Dame Jocelyn Bell Burnell at a science festival near Oxford. Jocelyn was the first person to observe pulsars. Pulsars are spherical, compact, highly magnetized and rapidly rotating neutron stars. They are formed when a massive star collapses. They appear to blink at a regular rhythm and were briefly called LGMs, or Little Green Man signals, as the regularity of the signal led to brief conjecture that this could be communication from intelligent life. Seven years later this discovery would lead to a Nobel Prize, though Jocelyn was not one of the recipients, and whilst many have considered her being overlooked as deeply unjust, she seems unruffled. She told me the disadvantage of getting a Nobel Prize is that they stop giving you other prizes, so she would have missed out on all those possible award dinners that have followed. At a press conference about pulsars – despite her being the first observer of them – the press concentrated on the older, tweedier male scientists in the room. Jocelyn recalls that she was only asked one question: a tabloid newspaper wished to know, 'Are you taller than Princess Margaret?'

Jocelyn is a pioneering astrophysicist, a strenuous campaigner for women in STEM (Science, Technology, Engineering and Mathematics) and the author of *A Quaker Astronomer Reflects: Can a Scientist Also Be Religious?* During her appearance on Radio 4's *Desert Island Discs*, Jocelyn said, 'There are scientists that are thoroughly anti-God, but equally I think there are scientists who try too hard to reconcile science and religion. I think there are pitfalls both ways. Of course I believe I am walking the perfect tightrope.'

Fred Hoyle's book *Frontiers of Astronomy* inspired her career. Jocelyn had just been learning about circular motion at school and, as she read about the motion of galaxies in Hoyle's book, she had her 'wow moment' and thought, 'This is a great subject and it seems to use physics, so I am going to be an astronomer; and then I finessed it, and it became a desire to be a radio astronomer.' Radio astronomy has the advantage over other forms of observational astronomy that you don't have to do the night-shift. Astronomy can be an isolating endeavour. *Dark Matter*, a poetry anthology that Jocelyn co-edited, includes Patric Dickinson's poem 'Jodrell Bank':

Now
We receive the blind codes
Of spaces beyond the span
Of our myths, and a long dead star
May only echo how
There are no loves nor gods
Men can invent to explain
How lonely all men are.

There can be communion with that solitude and darkness. When I walk around the observatory at Herstmonceux a few miles from the English south coast, in daylight and look at the dents on the floorboards where eager astronomers have been manoeuvring the telescopes over the years, I feel envy for those who possess the observatory at night. Not that working there at night is without its problems. In an ill-thought-out piece of daytime design, between the two observatories there is a lovely ornamental pond. Fine in sunlight, but a point of jeopardy for the benighted astronomer. When the pond was drained, numerous glass photographic plates containing the last vestiges of stars, obscured by pondweed, were found. The darkness of Herstmonceux is perfect for observing the stars, but less good for manoeuvring around decorative water features.

Jocelyn Bell Burnell has written that her Quaker bedfellow has had to bend to fit in with her astrophysicist bedfellow. Her faith has had to reshape itself to fit with what she has learnt about the universe, but the universe does not need to reciprocate in any way. She finds that dealing with vastness can be both daunting and dangerous. Daunting because of the scale, dangerous because 'You are working with this grand stuff and it can go to your head.'[3] She doesn't need a god to be an explanation for the universe. Her God is a personal god. This god is partly related to conscience, and to an acceptance of who one is – an acceptance of self. On occasions she sees this god as 'a sort of nourisher'.

Has her concept of God changed, with her changing knowledge of the universe? She doesn't think it has. She has not felt that she has to buy a package of religious beliefs, but that she can make up her own package, which accommodates all she knows of the universe. She can contemplate a god who is not all-powerful, a loving god who is not in control of the world.

The most likely place Jocelyn finds the numinous is still in the quiet contemplation of Quaker meetings, as well as in such activities as walking in mountains. She finds her transcendent moments when walking in the Alps. It gives her a sense of amazement, a sense of 'My goodness, is there really something like that out there?' Jocelyn has described the meeting of science and religion as curiosity mixing with anxiety.

Another place where curiosity often collides with anxiety – sometimes with great effect, and sometimes causing an unseemly unravelling – is in stand-up comedy. Most of the comedians I hang around with are eager to question things, to try to get lost in fascinations that may lead to a joke, before retreating into a paranoia that the punchline was not good enough. Comedians can often have all the curiosity of a scientist, but perhaps lack the attention span needed to crack the mysteries of the universe.

The Church of Quantum Physics

The comedian David Baddiel is devoutly godless, and it is only recently that he became interested in science. He was turned off science because he was brought up by a father who strenuously promoted it. David's scientific curiosity is now piqued enough for him to write plays exploring the ideas he is trying to understand. This is a useful way of using art; if you don't understand something, write a play about it or paint it, or turn it into a stand-up comedy routine. Punchlines can lead to enlightenment. *God's Dice* explores the themes of religion and quantum mechanics. David wondered: did belief in wave-particle duality amount to a religious-faith position, is this 'Bohr Again Christianity'?*

Baddiel describes himself as very Jewish, but with no belief in Judaism. He sees Judaism as incredibly abstract: 'An OCD way of controlling your life with 631 different tiny laws. It's not transcendent. It's very of this world and there's not even an afterlife, which is surely the whole point of religion.'

As he read about quantum physics, his inability to do the mathematics made him wonder if he was being asked to take a leap of faith. Were the priests of probability asking him to prostrate himself in front of their altar of equations? Was this language of physics like hearing the jabbering of Latin? (To be fair, I know very few physicists who will demand that you live your life according to the way of the electron, though on the downside that means you don't get complimentary bread and wine as part of the lecture.) For example, David's interpretation of the Copenhagen Interpretation of quantum physics is: 'We can't really explain what's happening at a subatomic level, but you just have to believe us.' He saw this as priestly physics with overlaps between mystery and religion and mystery and quantum physics, though he is quick to add, 'Even though the mystery of quantum physics obviously is real, and the one of religion is not.'

---

* Sometimes a pun is so niche, and has so few places to go, that you ignore the better advice of your editor and keep it in. Sorry!

His fascination with religion has also been stoked by one of his closest friends being devoutly religious. David is fascinated that his highly intelligent friend is also genuinely fearful of hellfire. *God's Dice* explains what happens 'when you get very intelligent people who believe the tenets of an organized religion somehow, due to very deep psychological crampons that have been put into them very early on.'

Quantum entanglement is the religious connection for the play's protagonist, Edie. Edie is both devoutly religious and a highly intelligent physics student. Nobel Prize-winner Frank Wilczek describes quantum entanglement as having 'an aura of glamorous mystery',[4] with links to many-worlds theory, but goes on to say that in the end these are ideas that should have 'down-to-earth meanings and concrete implications'. As with all the ideas that come from quantum physics, entanglement is something that can make your brain feel like it is physically vibrating in your skull and, if you think about it too hard when you first come across it, it can seem to turn your brain off and on again, in a desperate attempt to reboot you out of your confusion.

As I have come to understand it, an entangled system is one where, if you measure the state of one particle, you will instantaneously know the state of the particle it is entangled with, even if they are separated by millions of light years. Einstein called it 'spooky action from a distance'. It confounds many of us when we are first introduced to the idea. We are told that nothing can travel faster than the speed of light, and yet here is information seemingly travelling vast distances instantaneously. By measuring a particle in one place, you have changed the properties of another somewhere else – whatever the distance.

It is these ideas that allow all manner of New Age priests to sneak in their philosophies and sell their cures and self-help books, much to the consternation of the many physicists who protest that remarkable science has been used to sell snake oil and bunkum. The physicists will try to explain where the misunderstandings may lie, but as many of us can't get our heads around it in the first place, we cannot understand the misunderstanding, because we haven't reached the first point of understanding. And so we end up buying the miracle quantum skin lotion for our athlete's foot, in the belief that our toes will reach a superposition of smoothness anyway.

Edie's declaration to her physics tutor in *God's Dice* is, 'If I'm to believe quantum entanglement I may as well believe in God.' When David first read about quantum entanglement, he thought, 'This is miraculous.' He hopes to understand quantum mechanics before he dies. He is currently battling with quantum foam, which, if it exists, is about virtual particles that exist briefly as fleeting fluctuations in the fabric of spacetime. Part of the problem with much quantum physics may be that you have to think very hard to get any sense of understanding, and then it might turn out that you are attempting to comprehend something that doesn't exist in the first place. This is why I haven't yet started reading all those books I've got on string theory – I'm not sure I can think in one dimension.

David wants to be able to picture it in his head, but it is very hard to create a picture of something so abstract, from our daily experiences. He has moments when it feels as if a crack is opening and he can almost see it. He feels that those moments may be a bit like his religiously devout friend's transcendent moments with God. Like dreams, when you become conscious of them and start to try and track their plot in a woken state, the images you thought were in your head dissolve as you approach them.

Jousting with chaos, making out with Jesus

David's quest for enlightenment is not a quest for an alternative to a god, and all the ritual that goes with it. He does not hanker for ceremony, but some people find themselves troubled and lonely when they don't merely lose God, but lose all that structure and the socializing that goes with it. Is there a solution?

The Sunday Assembly was created in London by Pippa Evans and Sanderson Jones and soon spread around the world. It offers ritual, ceremony and a sense of belonging on a Sunday, for those who no longer have the glue of God to bind them. I was a little suspicious of the Sunday Assembly when I first heard about it. I think this came from a faulty instinct that free and secular thinking should be wary of congregations of any sort. It can seem that organized religion has staked a claim on the companionable things, ring-fencing the wonder, while allowing the secular to stay in the toolshed: 'Joy, wonder, purpose – we do all that – you get on with bashing away in isolation and leave the transcendent moments and the communal spirit to us. And no choral singing while you do it: that's ours too.' Ritual can seem so attached to mysticism that to behave in a ritualistic, ceremonial manner is like reneging on evidence-based thinking. I think some people in science can be overly wary of that area between astronomy and sun worship.

I have never missed the Church of my childhood. The pews were far too hard and the ceremony too tedious for me. Thank the Lord I never had a charismatic preacher, though I did have a quirky chaplain who sang a jaunty version of 'Yes, we have no bananas' while giving us tea and biscuits. I was lucky, because for many people leaving the Church can be traumatic.

Pippa Evans experienced the feeling of loss when the congregation that she ran with no longer fitted her ideas; when the rules started to rub her up the wrong way, such as when a man was thrown off the church committee because he lived with his girlfriend, or the offence that was taken because she was shoeless when she sang in the choir. It made her feel that the religion she was part of was more about the outer shell than about what was inside. Bureaucracy was more important than ritual – shared rules rather than shared experience.

When Sunday Assembly began, some attendees still carried damage from their church experiences, and Pippa and Sanderson needed to feel their way forward carefully. They didn't want to turn it into an atheist group, though some people pushed for this, as they still felt great anger from things the Church had done to them. The Sunday Assembly slogan began as 'Live Better, Help Often, Wonder More'.

Organized religion had been important to Pippa when she was a teenager. While her friends were doing sex and drugs, she says she was making out with Jesus. She was a regular at the Christian festival Soul Survivor. She now wonders how much of it was about Jesus and how much was about merely wanting to belong. She wanted 'a feeling', which, in shorthand, could be called God. This is something that fascinates me. What is the feeling of God? Is it the feeling I have when I am utterly beguiled by beauty, although for me that feeling is not converted into believing in a heavenly maker?

Now God was not in Pippa's universe, did she have moments when she thought, 'Oh, that sensation I just experienced was the one that I used to call God'? She remembers speaking in tongues. She doesn't know if she was really speaking in tongues, but what actually *is* speaking in tongues? I think I do it all the time when I walk around my house alone, speaking gobbledegook, opening the pressure valve in my skull to let out the mental kerfuffle.

Now the nearest Pippa gets to that freedom from the conformity of language is when she improvises onstage. She is a renowned improvisational performer and says there is a moment of creativity onstage – creativity that is part of interacting with a group of people, whereby there can be a sense of coexisting and fully being yourself. She says these moments have a kind of transcendence. 'We've been creating something onstage together and in that moment we've all known the same truth of that moment... the word "holy" comes to mind. But I don't mean a Christian holy, I mean something bigger and more beautiful than we could ever hope to create by ourselves. So it's something about a combination of human spirit together that creates a kind of alchemy, and that's what is holding you to me, I suppose.'

I can understand that. I am certain I have played to audiences who have imagined I am speaking in tongues, due to the speed and tangential nature of my jabbering. The most free that I feel is when I am onstage creating nonsense, which I am unaware of until it leaves my mouth. Loss of control in the right situation can be liberating, but just as speaking in tongues is best in the temple, so wildly improvising is better onstage than at a bus stop. The judgemental look of the queue can hamper any transcendental experience.

Pippa still believes we need to believe in something, and that we're all searching for what it is that gives us some connection to the bigger questions. She explains, 'So many people wanted to turn the Sunday Assembly into a safe space for atheists, which meant an angry space for people to be angry, rather than realizing that that's actually nothing to do with it. It's to do with humans. Humans hurt you.' In an early gathering there was a moment when the audience were asked to close their eyes for a minute in silence, which was considered too close to prayer for some. They have had to work their way around the different issues and conflicts that could arise, depending on the proximity of the idea to the Church tradition. Pippa believes many Churches lack 'that cheeky glint'. As a warm-up exercise in workshops, she asks people to look at each other and raise their eyebrows. It is the ice-breaker. It is a moment of connection through cheekiness and playfulness. If you are reading this on a train, do not do this exercise; but if you are reading this in a room where you know the occupants, give it a go.

Pippa still wishes she believed in God. Instead she sees herself jousting with chaos. 'We live in chaos, we've just found structures to make it look like we don't. In many ways God is chaos, and then I suppose another representation is the connected energy – energy between all life forms.' She wants to believe in something that's bigger than her. 'I can't believe that this is all just atoms and that we live and then we die, like there was something more poetic about it, you know?' She thinks she will still be wrestling with this until her old age, and she reckons that the Dalai Lama is still wrestling with it all too, but he gets to wrestle with a laughing Buddha, so it may seem like more fun.

'I am as irritated by the certainty of unbelievers as I am by the certainty of believers'

Pippa departed from the Church as a teenager. Richard Holloway left the Church having been Bishop of Edinburgh. His version of singing barefoot in the choir was performing marriage ceremonies for same-sex couples from 1972 onwards. He is my favourite former bishop (if the Bishop of Leeds retires before this book is published, I would like to make it clear that my league table of 'favourite former bishops' was drawn up whilst he was still a bishop, and when he retires I will redraw the whole thing). Over his long career in the Church, Richard's growing problems were around its ethics, morals and attitudes. The universe he lives in now is meaningless, but he tries to shame the universe by living as meaningful a life as possible. 'You don't solve the problems of suffering with a theory. You roll your sleeves up and do something to help the sufferers. It engaged me in a social-action ministry, do something for the sufferers, at least some good would have been done, maybe even a protest against the meaninglessness of the universe.'

The Astronomer Royal Martin Rees has described himself as a practising non-believer, which is where Richard Holloway finds himself now. He is comfortable with that paradox. 'If your faith causes you to be cruel, then I do not want to be part of it.' He has a moral revulsion that certain beliefs held in certain ways cause a cruel reaction in others. He doesn't mind what you believe, as long as it provokes kindness and understanding in you. He has always been very aware of pain. There is a tragic element to his philosophy, and the sense of the suffering of other creatures has always been part of it. 'We should be less certain about our certainties. Their empirical verification is the quality of the compassion they produce in us towards others,' he tells me. His position is active agnosticism – a commitment to unknowing. He tells me about the Caledonian Antisyzygies (which is now your new best Scrabble word, as long as someone already has 'ant' on the board. A long shot, I know). This is the presence of two polarities within an individual. Richard feels as if he is both an atheist and a theist, in a state of disjunction living on a boundary.

The universe has got bigger since he was a young priest – God perhaps seemed not so far away then. Now he engages with the universe being fourteen billion years old. He still finds it almost impossible to believe that something came out of nothing, though he is fascinated by how absence can become a presence. Driving home through the countryside one night, listening to Holst's *The Planets*, Richard was suddenly overcome. With the windows open, he started to shout, 'YES! YES!' while punching the air.

On another occasion, while walking through London, the world suddenly became a ballet, 'a feeling of being enclosed in the universe, the universe as benign to us and affirming us and wanting us to join in the dance. There was nothing druidical about the experience, you can't prove anything from it, but it was a moment of communion – a moment of union with everything as it was. Nothing veridical in the experience, something personally affirming about it; enough interesting people have had these experiences to keep your mind open for there to be more to the universe than you have given it credit for.'

One of Richard's favourite poems is 'The More Loving One' by W. H. Auden, from which he takes that the universe is without purpose or agency, but that it is where we discovered love.

'I am much more puzzled by religion than I am by science'

Victor Stock, the former Dean of Guildford Cathedral, is definitely top of my 'favourite former deans' wall-chart (thank heavens he was never a bishop!). Victor socializes with the Queen and still preaches sermons in Westminster Abbey and I am an atheist with republican tendencies, and yet we find ourselves agreeing on nearly everything. I asked him whether he saw personal conflict arising between his religious beliefs and the universe, as understood by the scientific method, and he said that he didn't find a conflict, though he added, 'I do find a conflict, but that probably comes from ignorance and not understanding the scientific language.'

In old age, he feels he has discarded a few of his religious certainties. He considers the doctrine of the Fall of Man to be a terrible mistake. This is certainly where my detachment from Christian religion would begin, if I wasn't already detached. This is the angry, abusive parent God, the one who tells you all the sacrifices they made for you, and how you have failed them.

'I made it all nice, but you had to talk to that serpent and ruin it, didn't you?' And then He chucks you out of the Garden, filled with questions about why His omniscience didn't work out the problems with that fruit tree that He put in the Garden in the first place, and why He made that bloody snake. He's not a good role model for parents.

Victor's engagement with his beliefs is always questioning. He does not want to be given certainties, and believes that the interesting clerics do not preach definitive realities. Maybe that is what people want but, from his experience, he believes that if you encourage people to be thoughtful, they will be thoughtful. 'You can seduce people into certainty,' he says, 'but I have never bought this idea that simple people can be patronized by telling them garbage because it is all they can manage.'

With someone like Victor, I always wonder what the difference is, for me and for him. There is so much common ground, but I am living in a meaningless universe and he's living in a meaningful universe. Maybe it is easier for me? By hook, crook, emotion and whatever knowledge I have, I have bumped my head against a universe without a grand purpose, and I simply have to work out ways of getting used to it.

How does that purposelessness measure up, if there is still a god in it somewhere? Victor says the meaning of the universe doesn't work for him, and it is all Billy Graham's fault. Billy Graham was ranked as one of America's eight richest pastors – and that means you are pretty wealthy; wealthy enough to pay for construction workers to widen the eye of any needle. As a young man, Victor saw a Billy Graham promotional film. It showed the universe, and a big sexy American voiceover said,

'Gazing up at the panoply of nature, the stars in their courses, who cannot but fall on their knees and say, "There is a Creator God"?' And Victor thought, 'Well, not me.' That sort of evangelical proselytizing doesn't work for him.

He questions whether the meaninglessness in the universe – the waste, horror and cruelty – outweighs the love, goodness and mercy. He thinks there is hope, but he is tentative. Victor may be as doubtful about the meaning of the universe as I am. He thinks the only difference may be summed up by a quote from Dante's *The Divine Comedy*: 'God is the love that moves the sun and the other stars.' He is content to live with what he considers to be ambiguous hopefulness.

I realize there is a bias in the religious people I meet. The fundamentalists and I don't drink together so much, so I am fascinated by the active doubt of the faithful that I know. I also want to know what they experience, when they experience God. For most of them, their god is far from literal and almost impossible to put into words. It is a sensation. Is it the same sensation I feel during a transcendent moment? When we are lost in beauty, those moments when our present self, quite unnoticed, disconnects and, on reconnecting, we could feel as if for a moment there was no divide between us and the rest of the universe – all our atoms vibrating at the same speed and interchangeable – is that when they feel their God? I don't believe I experience it any less. The elation can be in me, too, but for them somewhere in the translation it is interpreted as God.

I am fortunate to have no God-shaped hole in my life; whatever indentation he made on me when I was small, it healed over very quickly. Like so many questions, the existence of God can be refuted quite quickly, based on evidence alone, but the interesting question is not whether God exists; it is why some of us need Him to exist while others don't, and then what God actually is in the minds of so many different believers.

Tell people there's an invisible man in the sky who created the universe, and the vast majority will believe you. Tell them the paint is wet, and they have to touch it to be sure.

GEORGE CARLIN

CHAPTER 3

Armchair Time-Travel – Putting Out Your Beach Blanket on the Sands of Time

We are the guardians of forever. Only on this world do the million pulse points of time and space merge. Only here do the flux lines of forever meet. Only here can exist a gateway to the past.

The Guardians of Forever, *Star Trek*, 'The City on the Edge of Forever' (1967)[1]

As doubt and uncertainty increase, with attentive study of the universe, even the most solid and sure things that you can set your watch by become malleable and slippery. This includes time itself. Our developing knowledge of time's mutability, its entwining with space – like many revolutionary ideas – can seem alienating, but if we keep digging, then new comforts can be found (and you don't need to understand the equations*).

---

* You can presume throughout this book that I don't understand the equations.

Science fiction has had fun with time in many forms. I was brought up on the time-travellers of *Doctor Who*, H. G. Wells's *The Time Machine* and *Back to the Future*. That these stories relied on breaking the laws of physics was of no concern to me. The current understanding of spacetime is that you can change the speed at which you travel into the future, but you can never go back. The past is another country and its border controls are very, very strict. My favourite time-travel movie is *Time Bandits*, the product of the imagination of two Pythons, Terry Gilliam and Michael Palin, and with a cast that includes Sean Connery as Agamemnon. Six short and lowly workers for the Supreme Being steal the map that shows the holes in spacetime and attempt to plunder a variety of historical locations. Perhaps this doesn't break the laws of physics as, being workers for this Supreme Being, they may exist outside the laws of physics?

Time has been defined as the thing that stops everything happening at once. This summary has often been attributed to John Archibald Wheeler, an inspirational figure in the field of theoretical physics, although – like many great scientific ideas – it may first have appeared in pulp fiction. Here it is in Ray Cummings's story, *The Girl in the Golden Atom*:[2]

The Big Business Man smiled. '*Time*,' he said, '*is what keeps everything from happening at once.*'

'Very clever,' said the Chemist, laughing.

Our view of time passing is created by our sense of memory and panic. It is nostalgia for what is behind us, and anxiety about

My first photoshoot.

a future that is coming towards us too fast. As I write, I am sitting in a room surrounded by signifiers of the time, which feels as if it has both passed me and dragged me along with it. In a frame is the first photograph of me, fresh to the world, with my mother and my sisters looking down at me happily.

Fifteen minutes after the photo was taken, my eldest sister would bound across the village green to her school and tell everyone, 'I've got a baby brother!' Next to that frame, we are all together again, and that baby brother is now looking the oldest and greyest, despite being the youngest; mother no longer in the picture, having left the scene a few years before. There's my wedding photograph: my hair still brown and apparent then. There are faded photos of grandparents on the bookshelves. And I know that when I am gone, I might be fondly framed in my son's home one day.

On my chin I can feel the scar where I fell hard on the wall of the sandpit when I was three years old, and on my hand the scar from where I fell into an elderly rose bush while playing kiss-chase with the Baptist minister's daughter when I was four. One of the markers of the passing of time is entropy: the idea that as time moves forward, the universe becomes increasingly disordered. Looking at our scars and wrinkles and comparing ourselves now to the pictures of our chubby youth, that entropy can seem very apparent. Each memory, each accident, each stumble into a rose bush is another notch left by time.

There is no escaping the experience of time, and our sense that it keeps moving us forward. Despite the best efforts of both Dr Emmett Brown and the Gallifreyans, it is still a one-way experience. The forward momentum of time seems to be common sense; if it starts going the other way, it will play havoc with the cemeteries and with the invention of fire (not to mention Leonardo da Vinci's factory of *Mona Lisa*s[3]).

Few things pick away at the notion of scientific common sense more than the discovery that time, as we experience it via the ticking of a clock, is not as straightforward – or even as straight – as it may seem. The average existential anxiety arising from our path from cradle to grave becomes exotic and arcane, and perhaps more bearable, when we start to engage with ideas of time, whether by pondering the ramifications of Einstein's universe or simply picking up a pebble on a beach.

There are huge differences in our psychological experiences of any given minute, even for people sitting together in the same room. The minute that your daughter experiences as she plays *Minecraft* is perhaps so much faster than the minute you experience as you wait for her to put the game down and go and brush her teeth, as you have told her to three times already. Sitting in a cinema watching the latest superhero film, it races by for those in cosplay and drags interminably for those who really wanted to see that elegiac movie that won the Silver Bear at the Berlin International Film Festival about an old man repairing his late wife's boathouse in Sicily (he dies at the end).

How do we make sense of time: the time we're in, the time behind us and the time ahead? When I spoke to the physicist Jon Butterworth, he resorted to quoting St Bede:

The present life of man upon Earth, O King, seems to me in comparison with that time which is unknown to us like the swift flight of a sparrow through the mead-hall where you sit at supper in winter, with your Ealdormen and thanes, while the fire blazes in the midst and the hall is warmed, but the wintry storms of rain or snow are raging abroad. The sparrow, flying in at one door and immediately out at another, whilst he is within, is safe from the wintry tempest, but after a short space of fair weather, he immediately vanishes out of your sight, passing from winter to winter again. So this life of man appears for a little while, but of what is to follow or what went before we know nothing at all.[4]

There is something rather delightful when a physicist says, 'Sure, I've got a whole load of equations I can show you, but sometimes it can be best to turn to a seventh-century Benedictine monk. You can trust him, he's venerable.'

Merlin memories and other stone tapes

My last journey before the first pandemic lockdown was to Stornoway in the Outer Hebrides of Scotland. I was fortunate to be there on the only sunny day that the island had that February. The trip was precarious, as high winds and storms were imminent. A short drive from Stornoway are the Standing Stones of Callanish. They are 5,000 years old and, like other Neolithic stone circles, there is much debate about what their purpose was. Gail Higginbottom from the University of Adelaide believes they were built as a way of understanding the specific cycles of the Sun and Moon: 'It showed that their understanding of the Universe was that it was cyclic and made up of opposites. Dark and light, north and south, night and day.'[5] Others dismiss such interpretations as overly astronomical and too deeply mathematical.

My illiterate experience of the Standing Stones was of a dizzying and elevating connection with time. Unlike Stonehenge, you can touch these stones and walk among them. Something pulsed through me, when I placed the palm of my hand flat against a standing stone. It was a psychosomatic experience of time – a detachment from conscious reason and thought – a sensation of utter delight that collapses reality and creates a presence that is not anchored to the present date and time. It was a transcendent experience, which some would define as mystical. It felt like a connection to time, and yet also free from time. It was the sensation created by a quiet thought that said, 'These stones were here long before you existed and they will be here long after you have gone.' Attached to that thought is another one that says, 'Many have stood where you are now and contemplated the same thing, and rather than create an anxiety that you are such a small dot on the map of time, it offers the consolation that you are neither the beginning nor the end.' You are standing in others' footprints and, some day, others will stand in yours; in fact if Stornoway gets another sunny day in winter, someone might be standing in mine tomorrow. You are a coordinate on the map of spacetime, and briefly you imagine a view that is not attached to now.

When we put a hand on such stones, heavy and dense with myth, there might not be anything in a causally deterministic universe that is actually happening, but that doesn't mean that I'm not activating something. The knowledge of the history of myth can have an impact upon us. People talk about the stones emitting power and energy, but perhaps *we* emit the power and the energy – from our imagination – and the stones just echo what we have emitted. We project our stories and experience into them and then feel the pulse from the echo that comes back. The 'magic' in such an experience increases with the distance of time and the gathering of stories, and is perhaps increased by our uncertainty around what these stones were originally used for.

Improved technology for recording moments of time means that the past no longer looks like the past. The photographs of my grandparents are monochrome. Primary-school teachers find themselves having to explain to the children that although images of the First World War are in black and white, the world actually existed in colour back then. Before the camera, the past becomes even less literal, with its paintings of rich merchants, nobles, crucifixions and occasionally some peasants messily eating curd in that awful way that peasants are always eating curd. Here is where ancientness creeps in. These people are not like us. Nowadays I eat my curd with a knife and fork, as I am sure you do, too. Go back as far as the Standing Stones of Callanish and all is alien. We only have to go back a few thousand years from where we stand now and we call it 'ancient', leaving us short of words to describe the rest of the 4.6 billion years before the world was ancient.

This semantic problem – that what we call ancient is actually quite recent, terrestrially and cosmologically – is a stumbling block for many when trying to understand ideas that really take time, such as continental drift, or evolution by natural selection. We fail to see striking changes in the evolution of a kangaroo in the space of 100 years, and so evolution-deniers will offer up the kangaroo's failure to develop wings or a Velcroed pouch within a century as proof that change in nature doesn't act fast enough to give us the variety of living things that we have now.

Before we get to the physics of time as a human experience, we should perhaps acquaint ourselves with time and geology. I was brought up in the Chiltern Hills. On every woodland walk, you would see rabbit burrows and badger setts that exposed the chalk. We would dig into the disturbed chalk at the upturned trunks of fallen trees to make our dens, quite unaware that all that chalk was a product of living things – fossils created by plankton-like algae. Each handful of childhood chalk, perfect for the coy vandal to scrawl temporary graffiti on garage walls, was filled with tiny fossils. The chalk landscape is evidence of the warm seas that covered the surroundings of my now-inland childhood playground. A landscape of deep time surrounds us, a connection that sometimes you can reach down and scoop up in the palm of your hand.

The quick-quick-slow of geology

When Chris Jackson was growing up, he believed that his family might have been the only black caravanners in the Midlands. His holidays exploring the Peak District were an important part of the way he developed a love of rocks. Chris is now a professor of geology. When he explains geology to people, he can see a confusion with archaeology, because of those ideas that we have of what it is to be ancient. What helps Chris is being able to explain that 350 million years ago the Earth, in many ways, looked very similar to plenty of places on Earth now, and that to picture the past we can often picture the present. Such work is based on the principle of uniformitarianism, whereby we understand that the laws and actions that govern the universe have remained constant throughout time.

'If you look at what's on Earth in the present day, you can understand processes in geological time, because they haven't changed for hundreds and hundreds of millions of years,' Chris explains as he expands on the temporal dimension of geology. He likes to take people to look at layers of rock. Find an exposure of rock and all of the layers are from different years, covering thousands of years of accumulation and change. At a big cliff face, Chris looks at the rocks as if reading a book: he reads into them. Look at the exposed rocks on the Jurassic Coast of southern England and you can find pieces of rock that are 160 million years old, and within them you will see structures and fossils.

Rocks don't seem to possess the glamour of a supernova or the seahorse genome, but when Chris holds up a piece of rock and starts to tell you about the processes that led to its solidity and the billions of years it has existed on Earth, then that piece of rock is transformed into a story. He might even pop it into his mouth and give it a light chew, to get a sense of it.* Long before living creatures could leave fossil evidence, the rocks around us were the evidence of our changing Earth.

On Chris's childhood holidays, camping and walking in the Peak District of Derbyshire – in particular in the Hope Valley around Castleton – he first became properly acquainted with rocks of the Carboniferous Period that were between fifty and seventy million years old. During this period, across north Derbyshire and large parts of the North of England, giant coral reefs built up in the deep muddy basin. This basin has been revealed in the Hope Valley, so Chris would stand in the valley with his feet on Carboniferous mudstone, then look up above Castleton village and there he would see a big hill that goes up to Winnats Pass, and the hill that rises up is the front of the Carboniferous coral reefs. Inside the caves of the region you can see carbonates and all the wild creatures that built this coral reef.

These are the moments of standing still and time-travelling in your head, of feeling the sea rising, life changing, the undulations that seem so solid beneath you sinking and shifting, just as they did before, and no doubt will again when we are long gone. It is all still

* This is not solely Chris's idiosyncratic behaviour. I have spoken to other geologists who have explained that a brief rock-chew can give you useful information, but do it carefully – not near a sewage outlet – and don't bite down hard.

happening, invisible to our eye. Even the most lumpen, intransigent landscape is in constant motion; whether a glass of water, a mountain or a galaxy, nothing in the universe is motionless.

Mysterious fossils have been found in many places where it seems difficult to believe they ever should be. In 1987 an expedition to southern Chile by the American Museum of Natural History found fossils of whales in the Andes.[6] What was a whale doing on a mountain? Whale bones have been found at altitudes of more than 5,000 feet. Twenty million years ago this was the ocean floor. This is the work of plate tectonics, where a violent upthrust has turned the ocean floor into mountainside. This is what happens if you give geology a little time. Eternal cartographers, frustrated by the changes to their work required by such movement, will be happy to know that Quiming Cheng, a mathematical geoscientist, has predicted the end of plate tectonics in about 1.45 billion years, because by then the Earth's magma will have cooled.[7] Fossils will have to discover new ways to surprise whatever curious creatures are exploring then.

Chris returns to the Hope Valley a couple of times a year. He still cannot believe that he is basically standing in deep water in the Carboniferous Period, looking up at a coral reef towering above him. And north of Hope Valley there is Mam Tor. The rocks there are late Carboniferous, and the coal seams of England are found in them. You can pull whole leaves out of these coal seams that are hundreds of millions of years old. This is where Chris experiences his transcendent moments, but at such times you do not need to be a geologist to feel a connection with deep time. A landscape is beautiful without any knowledge – the slopes, the sky, a murmuration of birds – but with a knowledge of the immense amount of time needed to shape it, and of the processes required, the beauty becomes more than meadow-deep.

Chris can pick up a rock and see mineral assemblages, and what geologists call 'bed forms'. It can take him to a moment – hundreds of millions of years ago – when a river burst its banks and started spreading its sediments across the plains, beginning to make giant dunes. He can feel as if he is seeing it happen, and he tells the story in his head that he sees encoded in the rocks. Geology creates a very physical relationship with time. So whilst standing in a stone circle can create that sort of connection with the rocks, and with our own human history, if a stone catches your eye while you are wandering through a ploughed field or along a beach, then hold it in your hand and daydream your own geological time-machine. It is daunting to see how many potential stories of our past lie in the soil beneath us, and how we are increasingly able to jemmy out these stories. One day in the far future the intrigued geologist, from whatever curious species walks the Earth then, may find a piece of what was once you, then start to put together your story, and you will be able to tell them about what has been.

Space archaeology and the Ozymandias effect

Once, when I was walking through the streets of Cambridge on a dark and silent Sunday night, my mind detached from immediate physical reality and I saw a street laden with professors and students on bicycles, as if I was seeing the overlaying of many moments on that stretch of road over the last 150 years. I knew this vision began in my head, before even the faintest hint of haunting could form, but it was a beguiling vision of multiple past times. It was like a palimpsest – a repeated overwriting of moments of time. This palimpsest vision of time is something that Sarah Parcak deals with in the most concrete of ways.

Sarah is a space archaeologist. *Raiders of the Lost Ark* was a major influence on her career choice. She does not travel into space for her archaeology. Instead she is exploring the Earth – not seeking civilizations that might have sculpted faces on Mars, or painted on the cave walls of Titan, but looking at human history via images taken from space. In studying our past, she aims to find out about the origins of our resilience and creativity, to give us some hope for the future. We have a habit of believing that we live in exceptional times, but from her experience, political incompetence, boredom, lust and limericks are found in every time where humans have left their mark. Behind every pharaoh were enclaves of intellectuals and idiots. Underneath every pharaoh are all the poor slaves, wives and concubines that were buried with them.

Sarah uses data and photos taken from aeroplanes, drones, satellites and from the International Space Station. She looks for patterns on the surface of the Earth that suggest the forgotten structures of civilizations. It is like a space-based CT scan. Many traces of structures are partially or completely buried by soil, vegetation or under modern towns. The satellites enable her to look at different parts of the light spectrum that are invisible to us. When things are buried under the ground, they affect the overlying soils, vegetation and sands in ways that cannot be detected in the visible part of the light spectrum. For instance, an Iron Age ditch hidden beneath the ground can be observed, as the dense, moist soil will affect the overlying vegetation in such a way that it is going to be healthier. In the near-infrared part of the light spectrum, which is the part where you can see vegetation health, it is going to show up as being much clearer. Zoom out and zoom out and zoom out and you will be able to see the shape of an Iron Age ditch. During a recent long drought in England there were hundreds of archaeological sites that started showing up in places where archaeologists hadn't seen them before.

As well as using the latest technology to bridge time, Sarah also uses rude poetry. 'One of my favourite things to do, if I'm teaching an archaeology class around Valentine's Day, is to read ancient Egyptian love poetry, and it will turn your ears because it's filthy. It's incredibly erotic. People were very sexual. They talked about having their hearts broken, about cheating husbands – you know, the same stuff we read today. It's glorious.' Obviously your lustful poetry needs to be particularly potent if the target of your love always has in the back of her mind, 'Is this really the man I want to be buried alive with?'

Just as the laws of the universe lead to the principle of uniformitarianism, so the principles of human lust and jealousy show that being human throughout time has not been so very different, simply because we didn't have smartphones and sandwich toasters. Just as we can dehumanize people due to their geography or marginal physical differences, we can also dehumanize people through the distance of time. 'I've got nothing in common with you – you're from the Bronze Age.' Should an anomaly in the fabric of spacetime ever lead to an influx of refugees from the Neolithic period, I am sure the tabloid press will find all manner of colourful ways of seeding division.

The Nebamun reliefs, showing the life of the scribe
and 'grain accountant'.

When Sarah first saw the Nebamun reliefs – paintings from
the tomb of Nebamun, from around 1350 BC, rediscovered in
1820 – in the British Museum, she felt great tension in celebrating
life amidst death. She says, 'The rift in time was never so open,
as I stood there. It is probably the first time in my life when I
really understood ancient Egyptian tombs and what that meant,
to firmly embrace all that life gives us, while understanding the
impermanence of our existence.' Working on many tombs, Sarah
sees that the Egyptians knew that existence was tenuous at best,
but that they continued on with their daily lives. She proposed to
her husband at the Cairo Museum in front of the statues of Prince

Rahotep and his wife Nofret. 'If I was going to ask him to spend the rest of his life with me, then I felt I should ask him in front of two people who had pledged to be together for eternity. These symbols endure because they are mirrors.'

As Chris sees the overlaying of geological time, Sarah experiences the layering of archaeological time. Landscapes and townscapes are – like my vision on the streets of Cambridge – a palimpsest. In films about mummies that return to vengeful life or the unearthing of an ancient Satan, we are used to seeing the archaeological dig as a place in the middle of nowhere, far from towns and coffee chains, but many towns and villages were not abandoned in antiquity. People have lived in the same places for thousands of years and, even more than in Cambridge, this is very much so in Egypt. Sarah finds a great sense of continuity: under the modern buildings are the mounds of what was once there.

In Egyptian Coptic Christian communities the church is right in the middle of the town. Church staff will show Sarah the oldest part of the church, and she will ask them when they last did renovations. 'Oh, last year' will be the casual response. 'We were doing the plumbing and we found bits of an Egyptian temple.' Old mosques in the Middle East often sit on the sites of Christian churches, and they themselves used bricks from a Roman temple. There is a sense of great continuity. Sacred spaces remain sacred spaces, even as the gods change.

There is an Ozymandias effect, whereby archaeology perpetually underlines our impermanence. Since Sarah started doing larger-scale landscape analysis with satellite imagery, she has experienced 'the overview effect' – the cognitive shift in awareness reported by astronauts when they look down on the

Earth. Through the images that Sarah examines, her sense of the wholeness of civilization increases. She sees the connectivity, and yet the fragility, of these ancient sites, which has increased the rawness of emotion that she feels when she is at a dig.

'Getting older has made me rooted and in the present. Every time I go to a museum or a dig, I get very emotional now because I feel the people more. When I see the skeleton of a child, I can't stand it any more,' she says. Archaeology rehumanizes, or brings to life, the reality of those who lived in the past. It reattaches us, and we start to see that we are no more exceptional than those who left behind their love poetry, their insults, poverty, pain or mummified bodies. Sarah sums it up as, 'Never are we so present as when we are in the presence of the great past.'

Our roots are up above, beyond the light

There is a hazy line between where the past goes from being historical to being scientific. David Christian is a historian who has come to specialize in what has become known as Big History. He believes that if you want to teach the history of humanity properly, then you probably need to go back beyond human existence and teach the history of everything. He says, 'You can't just begin by announcing that there are human beings – you have to know how human beings came about.' So that would lead to the study of evolution, but to do that seriously, you have to go back beyond the apes, before the mammals, before the first fish that walked on a beach, to the origins of life on Earth. And why stop there? Is how the Sun formed part of history? In his pursuit of truly knowing and understanding,

David was pushed further and further back to the primeval atom and the beginning of everything.

He believes that if you want to understand the present, it makes sense to go back fourteen billion years. Can we integrate the historian's story with the anthropologist's story, with the biologist's story, with the palaeontologist's story, with the cosmologist's story? That is his challenge.

At the centre of the humanities should be the question: what does it mean to be human? David believes that many people in the humanities think that science is not terribly relevant to that question. It is not only in the mistakes of governments, and on the battlefield, that we can learn from history (not that we seem to), but also from the stories of other species, of the dead ends and the successful paths of other living creatures. Our history is not merely about us, and that connective tissue stretches back further than the Pyramids.

Patricio Guzmán's *Nostalgia for the Light* is a film that I often return to when contemplating ideas of a history that is bigger than us human beings. It is a beautiful and tragic documentary set in Chile's Atacama Desert, a place of history, archaeology, astronomy and struggles over human rights, yet a landscape that seems as dead as Mars. When you look up at the sky, you are looking back in time. Whether looking up or digging down, you are journeying into the past. You may be sure that the skeleton from the beast you have dug up is dead, so there is greater uncertainty when your eye is struck by the light of a distant star. Is that star dead now? What has been happening to it since the time when that photon of light was created via the nuclear reaction that turned hydrogen to helium?

The furthest star visible to the naked eye is 4,000 light years away, so the light you see now from that star was created at the same time as bronze was first being fashioned in China some 3,700 years ago.

Guzman's desert is full of interest and meaning. The telescopes of the desert survey the stars. They look back into time across the galaxy and, as they observe the old light, we can reflect on what was happening in that desert. The dryness of the desert preserves the remains of the dead, of mummies and miners, but also of murdered political prisoners. Look at the star system Tau Eridani or 47 Ursae Majoris and the light you see can take you back almost fifty years, to a time when General Pinochet was holding thousands who opposed him in the National Stadium. That light we see now comes from the time when the teacher, poet and singer Víctor Jara was murdered by the government – one of many. Those stars, in all their grandeur, with all their incredible power, can also act as markers for human atrocities: bookmarks in some of the more terrible pages of human action and injustice.

Guzmán shows us how astronomy and history can connect both brutally and beautifully. It was the first time I realized that astronomy was archaeology too. In archaeology the further down you go, the further back in time. In astronomy the further up you look, the further you see into the past. The nearest to the present is where you stand right now, and even that is not as it was. Panning across the desert floor, Guzmán talks of the meteorites buried beneath that can affect a compass, of the fossils of fish and molluscs buried in lakes of sand, but he concludes, 'I have always believed that our origins could be found in the ground buried beneath the soil or at the bottom of the sea, but now I think that our roots are up above beyond the light.' The night sky changed after I watched *Nostalgia for the Light*; this is what astronomy and its stories do to you. They transform the sky and they can transform you.

The past-present-future of always now

Before I can take on board the implications of the impossibility of truly living in the present, as my first contemplations are still a messy stew of thoughts, I find myself back with Professor Brian Cox. We are contemplating our next tour and, as he excitedly shows me through graphics and animations of black holes and the galaxies around them, I am casually confronted by the idea of time being an illusion. Sure, time feels to us like a moving thing, usually carrying us along too fast, but it might just be a great big block. The laws of the universe could severely affect the nostalgia industry, when it turns out that the past hasn't gone and the future is already here, if this idea of the block universe is true.

A few weeks before he died, Einstein famously wrote a letter to the widow of his friend, the engineer Michele Besso, whose revolutionary understanding of spacetime had not only changed the universe as we knew it, but had also found an idea that offered consolation in times of grief. 'Now he has departed from this strange world a little ahead of me. That means nothing. For us believing physicists the distinction between past, present, and future only has the meaning of an illusion, though a persistent one.' For Einstein, everything that happened *is* happening. The cosmologist Carlos Frenk sums up the situation as: 'You are your history.'

Here, Buddha, Einstein and Nietzsche all have something to say. Nietzsche wrote of the eternal recurrence of everything that happens happening for eternity. Few things are likely to encourage living a good life more than the realization that each dud date, spilt coffee or float in the Mediterranean Sea will happen an infinite number of times. Watch what you step in, as you'll be stepping in it for ever.

'I live in a world in my head in the world' – Paul Valéry

The idea that the past remains present is so seemingly fabulous at first sight that rather than turn to scientists, I turned to two of the most vivid imaginations working in fiction, both of whom have strikingly reimagined the possibility of comic books, and both of whom find inspiration in tales of myths and works of science.

Neil Gaiman's most famous comic-book work is the *Sandman* series. These are the stories of seven brothers and sisters who came into existence when time itself began. They are known as the Endless. As someone who plays with notions of time for a living, Neil is aware that there can seem to be very little in science that is comforting, and that the more you strip it down, the more nightmarish and Lovecraftian it becomes. There is a cold and pitiless universe that exists because of a few mathematical equations that care little about any of us, and this is particularly the case with ideas around time.

He sees the block-universe idea as one scientific formulation of the universe that does offer comfort. The block universe turns time into a slab – something that perpetually exists, and where we have merely travelled away from an event. The event is not obliterated by it being in what is our past. The existence of the home you woke up in does not stop when you depart for work, nor does the moment of time you departed from; the difference is that you can return to that home, but the laws of physics forbid you to go back to that moment in time. We take succour (if it was a good experience) in the knowledge that it is still there, and we can be glad that we cannot return to it (if it was unpleasant).

Alan Moore has thoroughly toyed with ideas of the block universe. He has always been playful with scientific concepts, from 2000 AD comics' *Future Shocks* via *Watchmen* to his vast epic *Jerusalem*. A personal favourite is *Jack B Quick*, the story of a boy genius whose messing with the laws of the universe leads to dreadful consequences. When his farmer father has cattle problems, Jack seeks solutions through spotting loopholes in the theory of relativity, which leads to creating a sun that eventually collapses into a black hole and yields a cow that can only be milked for X-rays. In another strip, Jack creates a Turing machine, which is little more than a wheelbarrow with a scarecrow, some junk and a tape recorder. Nevertheless the wheelbarrows take over the world. Alan sums it up, 'Eventually Jack realizes, when the whole world is enslaved by these scarecrows-and-junk in a wheelbarrow, that if we just stopped pushing the wheelbarrows... they'll be helpless!'[8] Within that short story is a pertinent criticism of the political world that we seem unable to imagine an alternative to.

It is in his vast novel *Jerusalem* that Alan explores the idea of the block universe most thoroughly, setting the book across centuries of history of Northampton, his home town. It is an intense psychogeographical exploration, in which he plays with notions of the structure of time. Alan came across the idea of the block universe reading Einstein's letter to Besso's widow, when he was halfway through writing his novel, a process that took many years. I happened to speak to Alan that day and he was particularly jocular. He was rather pleased he had found out that Einstein agreed with him. Though he had not come across the scientific idea of the block universe, the ideas within it were something he had been playing with in his imagination.

Alan's suspicion that time was more than it appeared to be began as a child, when he was looking at some faded sepia photographs of elderly relatives who had died before he was born. He couldn't actually put it into words, in that it was sensation rather than a thought. Looking at these old men in a pub garden somewhere, squinting towards the camera, he remembers thinking, 'Do these people know that they're dead?' He had a feeling about how photographs captured the light of a particular moment. It reminded me of what Alan has said about his mother's death and the discovery of an ancestor's birth caul.* Looking through the objects in his mother's house, he describes 'shoe boxes full of old light, war and buried weddings'. Since then, for me, every attic box or old drawer I open seems to contain that old light. I look again at the photo taken shortly after my birth: what do those figures in the photograph know?

'Way back in the past, somebody's face or something that somebody was doing gave me the first sense that the future was

---

* A birth caul is the membrane that, on rare occasions, can cover a baby when it is born. It would be preserved and was considered to bring good fortune to sailors.

already there and that the past still persisted,' says Alan. The first time he actually came across the idea that time may not be simply linear was in Kurt Vonnegut's *Slaughterhouse-Five*, in which Billy Pilgrim and the reader find themselves moving through time and space, back and forth, from a Second World War prison camp to a 1960s optometrist convention, to a zoo on the planet Tralfamadore. The Tralfamadorians experience reality in four dimensions, giving them access to the past, present and future.[9]

Throughout Alan's twenties this view of time became a gathering conviction and, once he understood Einstein's theory of relativity and what spacetime actually implied, his conception firmed up. Here was a 4D universe, where time was a measurement of distance. When researching his work about Jack the Ripper, *From Hell*, he came across late-nineteenth-century theorists writing about the fourth dimension and it occurred to him that what was being said was that spacetime was solid, and that whatever was part of that solid was eternal.

He sees a human as being projected through time like a kind of endlessly coiling centipede, with lots of arms and legs that wind all over the world and double back upon itself – everywhere that we have ever been. It is a beguiling and evocative image to ennoble your daydreams in the waiting room. 'At one end of it, there would be an ovum and some sperm; at the other end of it there would be cremated dust. But the bit in the middle, where we are alive and are conscious of being alive, that is as eternal as everything else in the universe presumably... this seemed to me to be perhaps the most startling implication of a four-dimensional universe that would employ an endless repetition. I thought I was the first person ever to come up with this idea. Then I found out about Gurdjieff, Joyce, Nietzsche...' For Alan, if time is a distance, or can be conceived of as a distance, then that removes an awful lot of existential worries, without actually appealing to the supernatural.

The first fan letter Alan received about *Jerusalem* was from a reader who told him he remembered the precise moment when he realized that everybody he knew would one day die. This sucker-punch of anxiety hit him as he sat on the sofa chewing a wine gum. He said that it opened up a big space for depression and anxiety, which blighted much of his life, despite seeing therapists and analysts, because they couldn't get around this central fact of mortality. Seeing time as expressed in the block-universe vision of Northampton, which Alan had created, gave him another way of looking at this relationship with what he had imagined to be the relentless arrow of time.

This relationship with time can create a feeling that every moment is wondrous and complex, an eternal gem in a crystalline atmosphere. All the moments are flies in amber, and all of our stories are going on simultaneously.

Alan has written other stories about time, including one about the first femtosecond of existence. The real problems with time for physicists seem to occur near its start. Alan's femtosecond goes on for ever because time has not yet reached the point of its smallest division. The story is called 'The Improbably Complex High-Energy Event'. His inspiration came about through his understanding of the implications of the concept of entropy. The idea is that the universe must have started in an improbably complex high-energy state that will eventually run all the way down to the complete disorganization that is suggestive of entropy. He has packed the first femtosecond of existence with all manner of events. Alan describes the story as totally unscientific, although he defies anybody to challenge him on it.

Losing the arrow of time

If anyone is going to challenge Alan Moore, it might be the cosmologist Carlos Frenk. I used to think that cosmologists had done very well in getting all the way back to an understanding of the beginning of the universe, apart from a pesky ten to the minus forty-three seconds, right at the start of the Big Bang, which seems like a very short time indeed and not really worth so much fuss. Then I spoke to Carlos and I discovered it might not be as simple as it appears. There is a slight problem with that first ten to the minus forty-three seconds if you want to consider it to be exactly that – ten to the minus forty-three seconds – because time is not that simple, at what was the beginning of time.

Carlos describes it as 'a very profound question that definitely makes things more interesting'. It is for that reason that he has an answer, which he believes is not necessarily satisfactory. It concerns the arrow of time, which distinguishes the past from the future. Most of us believe that the past was before now and that the future is still to come. We can never move into the past or the future because when we get there, it is always the present. The present is where we live, though as we saw with the block universe, moments are not gone or do not disappear, but simply become inaccessible.

We feel that time progresses in a direction. In another book, by someone cleverer, we could talk about the idea that time does

114

not really exist, but I am not the man for that job.* For now, let's just presume that there is an arrow of time, and go back into the past. Carlos explains to me that imagining the past – going back in time – is the equivalent to existing in a different state in which the density is much higher. Yesterday the universe was smaller than it is today, and every step back in time is a step back to a smaller everything. The further back you go, the higher the density and the higher the temperature.

By the time you get to the Big Bang, the density is incredibly high, as is the temperature, but the laws of physics still apply; 100,000 years after the Big Bang, this is no problem. One second after the Big Bang, still not a problem from the perspective of the laws of physics. Even at a millionth of a second, it all seems okay. It is just at ten to the minus forty-three seconds – Planck time (named after the German physicist Max Planck) – that things become a little lawless. Planck time is a unique time in the history of the universe. It is here that one of the great problems of physics becomes apparent. Gravity and quantum mechanics simply don't get on together, and this can be ignored for much of the time.

The nuclear forces that bind the nuclei together are ten-to-the-forty times stronger than gravity. Gravity is very weak, despite it still hurting if you drop a hammer on your foot. In nuclear physics, you don't have to care about gravity, because it is irrelevant. If you are going to study a nuclear explosion, the gravitational field is irrelevant. Carlos told me that if you want to work on cosmology, then you don't care about nuclear physics – 'you just worry about the whole universe'. When the two combine, which

---

* You'll want Carlo Rovelli's *The Order of Time* for that.

is at Planck time, they are equally important. Carlos blames Einstein for not living long enough to come up with an answer to this problem. The problem that cosmologists are trying to crack is quantum gravity – a theory that doesn't yet exist.

Here is where the possible reality of physics really confuses me. Carlos explains that the arrow of time has to do with entropy, and entropy in the universe has to do with gravity. 'As we wind the clock down and we enter this quantum-gravity period, the arrow of time gets confused because of these quantum fluctuations. When you get inside Planck time there is no longer the future or the past, so the arrow of time becomes random. And so the concept of time, in a sense, breaks down because if you no longer know what the future is, and what's the past, well, time is useless. There'll be no future and there would be no present. And so when you ask, "Then what happens? What happens to time?", I think time kind of self-destructs in the quantum-gravity regime; it kind of disappears. In a sense it doesn't disappear. But time in physics is just a coordinate.'

As I listen to Carlos, I am lost in incomprehension, but I don't feel uncomfortably ill at ease. There is something comforting about this realm where time and space and gravity and quantum mechanics lose their individual definitions. As his explanation continues, I feel much the same way as I did when I took too much codeine, after one of my many dental mishaps. I am disconnected and fuzzily warm. It is as if my brain knows that somewhere in its grey mess this all makes sense, but that sense of making sense will disintegrate if I scrutinize it too much.

Carlos continues, 'Time times Velocity equals Distance and it is the fourth dimension. In relativity it is nothing other than time multiplied by a very large number, which is the speed of light, which is so huge that it makes this dimension of time very, very big. One second corresponds to 300,000 kilometres, because the speed of light is at 300,000 kilobits per second. So one second, which is a kind of normal lapse of time, translates into distance... So that's why our perception of time is different from distance, because it's just so huge that we separate it mentally. Back now to the question of what happened to time towards the Big Bang – this huge dimension becomes sort of compactified. The unit is so dense that the separation of time from the other dimensions doesn't exist any more. I suspect that what happens there is everything gets mixed in. So you don't know what left and right is any more. You don't know what is the past and the future.'

In telling me this, Carlos had broken his own law. He made a rule, which is only to talk about physics. Quantum gravity doesn't exist, but now he was speculating. Don't think that I sat there nodding with instantaneous understanding, either. These are vast ideas about the universe and, when I read them back, my mind bounces back and forth between being lost and found and then lost again.

Carlos finishes it off with, 'So what I should have said, Robin, is: Don't ask me that question. Because we don't have quantum gravity. And until we have quantum gravity, I won't be able to answer you, because I have to leave my comfort zone, called the laws of physics. However, I did indulge and tell you my own view of what happened. As we approach the Big Bang with time, we don't lose it. We lose the arrow of time. There's no future, there's no past, it's all a mess. And it's all a fluctuating mess,' Carlos concludes, with another of those beautiful messes that we find the world becoming when we try to comprehend it all.

There is a reality where time is not time; where the universe, despite being so small at that point, no longer fits in our brain. As the codeine thins in my blood, I wonder how it can be considered that science disenchants the world. The power of the spells of physics to make the empty solid – and nothing, something – seem greater than anything I have ever read in folk tales.

Sometimes, with time, the problem is proving that it even exists. It is perhaps an unsettling thought to think that time is not real. We are unable to stop outside time to take a look at that idea. We are in it, even if it is not what we think it might be. The question of the reality of time is a problem that the theoretical physicist Fay Dowker has with her colleagues, when she tells them that time is not an illusion. For Fay, time is not merely a figment of our imaginations. I asked her if the block universe is a universe where every moment exists for eternity?

Fay replied with a quote from Omar Khayyám rather than from the Venerable Bede:

The Moving Finger writes; and, having writ,
Moves on: nor all thy Piety nor Wit
Shall lure it back to cancel half a Line,
Nor all thy Tears wash out a Word of it.[10]

Which she summarizes as, 'In one sense, yes. In another sense, no.'

Fay doesn't believe that we live those events for ever. She believes that we experience time passing and events happening. Once they have happened, we don't experience them again. She thinks that is a genuine, real, fundamental physical process, which doesn't correspond to anything in current physics. One of the questions – one that impinges on free will as much as any neuroscientific experiment – is: If the past is real in this block universe, is the future open or already set in place in the block of time? Is the life you are living on a single path? Fay thinks these questions about block or not block, about the passage of time being real or being an illusion, are logically distinct from the question of what the fundamental dynamics of the world are.

She says that you can conceive of a deterministic block-universe model that is not deterministic. This means that you know there is a selection of different universes that could happen, and it's a random choice that the universe makes as to which block is selected. There is a block view with determinism and a block view with randomness. You can have a view in which there is a process of coming into being in a deterministic way or in a non-deterministic way.

She explains the model of the random walk that offers us some freedom. There is a walker who can be at any one of a discrete set of positions on a line. There is a deterministic model of the walker, which states that the walker must do one thing or another at a particular time. There is a law that tells the walker which way to go, but there is still in the conception a process; the walker has to do it. It's not that the walk has happened and there is a real physical process of the walker taking a step to the left or a step to the right at each time. Then you can also consider a random walker. There is no definite law for which way the walker will move. You could just give them a probability about going to the right each time, but it is also a process. The walker really does take the step to the left or the right – the probability half one way and half the other way. That's random, in that you don't know which path the walker will take, but there are many different possibilities. It is not set out in a timeless way in a block.

Fay summarizes it as, 'You can have a deterministic process or you can have an end-deterministic process. So the two things that you know, they're sort of logically separate, right? So, yeah, they kind of impinge on each other, but know that you can have theoretical models which are blocky or not blocky. Which are deterministic or non-deterministic? These four different combinations are all possible.' It all seems so simple when I look at my wall calendar, yet what lies beneath it is so much more than just notes of a public holiday and dental appointments.

One of the many things that never entered my physics lessons, as we made pendulums swing back and forth, was how the laws of physics might impinge on my personal freedoms, and whether the very notion of personal freedom was an illusion.

When Fay presents her introductory lessons on this, she can see that her students have their minds blown. They find it very difficult to comprehend, and familiarity with such concepts doesn't come easily. After she has explained the four-dimensional view of the universe, some students ask her, 'Do you think this way about the world, when you're crossing the road?' Her answer is 'No'. She thinks of that bus as a thing that really exists, a three-dimensional solid object hurtling towards her. In road safety, she takes the Newtonian perspective quite seriously. But in her more contemplative moments, she returns to think of life and the universe in the Einsteinian way, because that is how she believes it really to be.

Fay believes that science has to connect with our experience. The world is as it is, and our experience of it may seem different, as our senses have evolved to perceive it in a certain way. Other species with other needs for survival sense it in another way. Science can't be in contradiction to our experiences. She thinks this is vital. 'If we can't make the theory coordinate with our experience, then it's a bad theory. We probably have to throw it away. The question is: how do we bring science and our scientific understanding into coordination with our experience? If she made a claim that the world can be understood in a certain way, then it must impinge on the way that we understand ourselves.

Fay sees Einstein's writings as both deeply scientific and deeply spiritual. He said there was no reason why the world should be comprehensible to us all and that it should be so ordered. This order was of deep wonder and mystery to Einstein. We can comprehend it in a subtle way, but we have to work really hard to understand it.

Perhaps we are not equipped to perceive time properly. It seems impossible to imagine nothing, and that is what we would have to do. There was a time when there was no before, and there will come a time where there is no after. We grew up in spacetime, in four dimensions: that is the view we have. Could we really imagine and believe it in any other way, whatever the equations may say?

We think that something in the past is not real any more, but it is still there in general relativity. It is part of the curved spacetime in which we live. Just because an event is behind us in our timeline does not make it any less real. When someone you love dies, or when a time that you valued is over, it can never be recaptured, but it is there. Inspirational speakers talk of us 'living in *the now*!', which, however well intentioned or financially motivated, is simply not feasible and possibly not as inspirational as it may seem.

One of the most-quoted movie death-scenes comes in *Blade Runner*. Rutger Hauer, as replicant Roy Batty, faces his certain death, robbed of a long life by his manufacturer. 'I've seen things you people wouldn't believe. Attack ships on fire off the shoulder of Orion. I watched C-beams glitter in the dark near the Tannhäuser Gate. All those moments will be lost in time, like tears in rain. Time to die.'

But maybe those moments are not lost. Some are trapped in the grand vision of the block universe, others may be trapped in a piece of rock or amber, the shape of a landscape or on the pages of a book. Our lives remain finite, but there is a sense of eternity in the laws of physics, too.

I stop and do nothing. Nothing happens. I am thinking about nothing. I listen to the passing of time.

CARLO ROVELLI

CHAPTER 4

Big, Isn't It? – On Coping with the Size of the Universe

The Brain – is wider than the Sky –
For – put them side by side –
The one the other will contain
With ease – and You – beside.

Emily Dickinson

One of my favourite instruments of torture is the Total Perspective Vortex created by Douglas Adams in *The Restaurant at the End of the Universe*. He writes, 'When you are put into the Vortex you are given just one momentary glimpse of the entire unimaginable infinity of creation, and somewhere in it a tiny little marker, a microscopic dot on a microscopic dot, which says "You are here".' This is close to the sort of torture you might experience during a Brian Cox lecture when you are confronted by the size of it all, the shock of this immensity only being partially softened by his soothing delivery.

I remember the first time I was terrified by the size of the universe. I was nine years old and I was in the bathroom. I have no idea what I had been watching or reading, but when I realized the possibility of an infinite universe, I nearly fell into the toilet. Imagining infinity played havoc with my middle ear. It was as if something without boundary of space or time rushed into my skull and tipped me over. The wall in front of me dissolved and I could see deep into the universe, and the faster I moved through it, the clearer it became that there was no end.

It is a memory that I can replay over and over again, and I can vividly feel that rapidly condensed sensation of being utterly lost and alone. Then the nightmares came. I would be on some wrought-iron factory stairs, like an M. C. Escher drawing made of Meccano, and then my mind's camera would start pulling back into the darkness and I would become smaller and smaller, until the horror of it all brought me back into a waking state. I would feel dizzy and confused, and this image would hound me for the whole school day.

If you are worried about the size of the universe today, remember that tomorrow is even more worrying as the universe will be bigger still. The magnitude of our understanding of it is increasing, but it is also literally expanding, second by second. In the human imagination the Earth used to be the centre of a little universe with a sun, the Moon and a few planets moving around it, and the twinkling stars just a decorative backdrop. Everything was a bit more local. Around 450 BC the Athenian philosopher Anaxagoras suggested that the stars might be like the Sun, and therefore much further away than presumed; and the Sun was also larger than had been imagined. But, like other astronomers after him, these contemplations led to prison.

There were other notorious astronomical proposals that led to punishment. Galileo's argument for the Copernican heliocentric model of our solar system, after he observed the moons of Jupiter, led to him spending the last few years of his life under house arrest. A few years beforehand, Giordano Bruno was less lucky with his punishment. His proposal that the Sun was a star, the universe was infinite and that there were many worlds led to him being burnt at the stake, though I do remember one astronomer telling me there were rumours that he had been annoying the churches for years and they were looking for any excuse to throw him on the pyre.*

Fortunately, astronomical proposals are less likely to be imprisonable or executable offences now, though I have heard custodial sentences might be introduced for people deliberately misunderstanding the holographic principle for financial gain. As the astronomer Priyamvada Natarajan points out in *Mapping the Heavens*, less than 100 years ago the Milky Way was the entire universe to us. The increasing power of telescopes changed the size of the universe, and by the 1920s indistinct nebulae of uncertain distance could be imagined to lie within our galaxy. By 1924 the Mount Wilson Observatory was able to demonstrate there were things beyond our galaxy. By 1929 not only were other galaxies observed, but they were seen to be moving away from us.

* Sorry not to recall which astronomy festival I heard this at – clearly this anecdote must be filed under 'Tittle-tattle'.

If the universe was getting bigger, then the questions were: How small had it once been, and what was its destiny? As the universe expands, more stars form. The star-formation rate is a remarkable 4,800 stars per second. I say 'remarkable', but of course it is merely remarkable to me or you because we may well not have contemplated such creativity before, and the night sky as we look at it seems a pretty stable and unchanging thing. To the universe, it is not remarkable at all, but merely the everyday, humdrum way things are in the cosmos. The size and dynamism of the universe seem shocking because our contemplations are normally about whether the milk in the fridge is on the turn or whether we should take our umbrella with us, rather than on star-formation rate, although I do know some people who spend more time contemplating star-formation rates, but they do often have runny noses and lumpy tea.

One of the first staggering statistics of the universe that spun my brain was that there are more stars than there are grains of sand on the Earth. Recently I discovered there is a problem with that statistic – not with the stars, but with the sand. It is easier to calculate the stars in the universe than the grains of sand on our planet. When I first read about the stars in our galaxy, there was a suggestion that there were up to 400 billion, but more recent books have suggested somewhere between 100 and 200 billion. It is tough to count stars on your fingers, so one of the main methods comes from estimating the mass of a galaxy, which is done by looking at how the galaxy rotates. That such calculations are possible staggers me as much as the results, but then again, the nocturnal dung beetle *Scarabaeus satyrus* uses the light of the Milky Way to navigate its balls of dung in the right direction, so maybe we shouldn't see as so exceptional our leap from astronomical navigation to where we are now.

It was estimated that there are as many galaxies as there are stars in our galaxy – so maybe 100 billion galaxies, with an average of 100 million stars, although more recently that has been upped to two trillion stars in the universe. There comes a point where millions or billions or trillions may have more meaning for the astrophysicist than for the casual lay person. Most ordinary brains reach a point at which mind-boggling vastness, and a billion here or there, doesn't change the sensation in the pit of your stomach. All the calculations and numbers could be replaced with that delightful opening narration from the classic film of love and hope in war, *A Matter of Life and Death*, when, as the camera journeys through the stars, the narrator nonchalantly says, 'This is the universe. Big, isn't it?'

One of the ways of comprehending the difference between an astronomer's mind and my mind, or yours, is to see what words like 'nearby' and 'local' mean to an astronomer. If a seaside Tourist Information Centre used the word 'nearby' in the way an astronomer does, you would definitely be writing an intemperate review on Tripadvisor. It is like visiting the TIC in Edinburgh and being told that features of local interest include the geysers of Enceladus and the canals on Mars. 'Near' and 'far' broaden their definition once they escape the pull of Earth's gravity. The nearest galaxy to the Milky Way is Andromeda, 2.5 million light years away; but don't worry, they'll probably collide in four billion years, so it will at least be a little more local then.

Beyond the boundaries

Though the size of the universe may be an astronomical issue, our sense of diminishment in all of this can be a psychological one. Mike Brearley was one of England's greatest cricket captains, losing only four of the thirty-one Tests he captained. You might be surprised that he is one of the first people I go to, when writing a chapter on the size of the universe, but he is also a psychotherapist and this is an anxiety issue. (I have yet to find the right international darts player to explain Gaussian symmetry, but I'm not giving up hope.)

I first met Michael in the small Welsh town (or large village) of Laugharne. Before the event I was hosting with him, he wanted to check that we wouldn't talk solely about cricket. I assured him that I would not be dragging him through the minutiae of every spun ball at Lord's, but instead we would get in one fabulous story about the eccentric batsman Derek Randall and then, while the audience felt reassured, we would leap straight into the psychology of William James and the novels of his brother, Henry.

I reckon we got at least thirty minutes in before some of the more ardent cricket fans cottoned on that the James brothers were not two forgotten Victorian cricket heroes. I love cricket as much as the next man – as long as the next man isn't most men.*

Michael remembers the first time he thought about the size of the universe. He had just found out that the Sun was bigger

---

* This is not to say that I don't like cricket, as it is in my top three favourite sports, and by stretching it out over five days, it is the nearest we get to a ball game invented by Ingmar Bergman. My other favourites are darts and snooker, because they have a Sergio Leone quality of lone individuals outstaring each other.

than the Earth. This terrified him. Until then he had imagined the universe as a box, and the things in the sky were painted on. How could the Sun be bigger than the ceiling, if that's all there was? If the Sun was bigger than the ceiling, Michael felt that something was going to smash the whole world to pieces. It would destroy the box that was the world as he had imagined it.

He was struck by a terror of the unknown, and his own insignificance, in Bognor Regis. His family was staying in a bungalow, and Mike was sleeping in the bedroom nearest to the road. He would lie awake, listening to the traffic, and sometimes a loud lorry would pass by. The lorry would come from one side to the other, and he would hear that change in sound, getting louder and louder, and then it would gradually tail away. As the lorry got louder, young Mike felt that time was getting slower and slower, and the terror was that everything was going to stop. Time was going to stop.

Now, looking back at these thoughts with the benefit of his analyst's education, he can try to understand that small child's fear. It comes from a dizzying sense of being lost in infinity – being lost in something you can't conceive of. This sort of anxiety does not necessarily ebb away at the end of childhood, as we continue to feel the echoes of childhood anxieties throughout our lives.

My own infinity anxiety increased; not only is there more than one size of infinity, but there is more than one size of infinity anxiety. It came from watching TV again, rather than on a trip to Bognor Regis. I saw *Powers of Ten*, the celebrated documentary short where the camera seemingly zooms out from a Chicago picnic scene, the distance increasing by a power of ten every ten seconds, getting further and further away from the man and the woman and their bowl of fruit. It described itself as 'A film dealing with the relative size of things in the universe and the effect of adding another zero'.

'This lonely scene, the galaxies like dust, is what most of space looks like. This emptiness is normal, the richness of our own neighbourhood is the exception,' says the narrator as our journey pauses 100 million light years away from Earth. Then we crash back down to the picnic at five times the speed we left, and find ourselves on the hand of a picnicker. Now the journey becomes a dive into the human body, with a reduction of 90 per cent every ten seconds, into blood cells, the cell nucleus, a coil of DNA, outer and inner electrons and finally to the nucleus of the atom. This back-and-forth journey still makes an impression forty-two years on from when I first saw it. The falling from the great height of 100 million light years away is still dizzying. To see what lies beyond us, around us and within us is important, when giving us a sense of place and a sense of rarity. The film gave me the scope – perhaps as it did for many children – to wonder if we could ever travel so fast that we could get to the end of the universe and then imagine what was on the other side.

One of the most-asked questions at any physics Q&A is: What is the universe expanding into? Our brain cannot imagine nothing, so we feel the need to place everything in something. But our universe *is* expanding into nothing. It is counter-instinctual, because our animal instincts are not adapted for astrophysics.

The universe is everything. It is all that exists. If there was something outside that it was expanding into, it couldn't be outside the universe, because that would be the universe, too. Any desire to travel to the end of the universe is hampered by the fact that the universe is expanding faster than the speed of light, and the universe has a speed limit that forbids travel faster than the speed of light – which is very canny for any universe with a secret to keep from us. The universe is allowed to break its own speed limit, but what is within it is not. On the size of the universe, Douglas Adams wrote, 'It is the lot. It is not nothing.' You are either in the universe or you do not exist, probably.

Almost everything that is in the universe will remain unexplored by us, at least for the time being. The good fortune we have is that curiosity and ingenuity have led to a set of laws that have, when combined with observations, started to allow us to work out what large areas of space should be like, and what they will contain, even if we can't ever reach them. The big picture is beginning to be filled in, whereas the small details, such as whether there is life on other planets, will be trickier to ascertain.

When I asked Brian Cox how he felt about the size of the universe, he looked down at a blade of grass. He knows that looking through a telescope at Saturn is a powerful experience, but he sees no difference between viewing Saturn or a planetary nebula and examining a blade of grass. You should be delighted to exist, and even happier that other things exist around you – they give you something to look at, and things to do. Brian believes that what the immense size of the universe should do is remind you that you are physically insignificant. We should not attach any great universal reason for our own existence that suggests any more importance than the existence of a pebble. Our existence is a cause for humility as well as for celebration.

The human/pebble comparison is Brian at his most hardcore. I think I caught him on a day when he had just had a new drive laid. He admits that he doesn't believe pebbles think, and he can see that there is something extremely special about us locally, because we bring meaning to the universe. But what does it mean to 'bring meaning' to the universe? 'It's a good line of questioning, because I think if you take that literally – that we bring meaning to the universe – then the thing you have to be most delighted about is your own perception of it. The fact that you exist, and think within it, is more remarkable than the stars.'

So, within the anxiety some experience of being lost in the universe is the upside that you are something that is capable of feeling lost in the universe; that you are able to contemplate more than your prey or your cave; that you can wonder if the universe has anything else on the other side; and that you can contemplate how long it might take to get there, if there is.

Brian sees the pessimistic reactions that people have on hearing about the size of the universe, or even just the size of our galaxy, as being attached to the worry that we are less important than we thought we were. With our worry about size and significance can also come the relief of being less important, too. Realizing the incredible weight of the universe should be a weight off our shoulders. Brian sees these as 'orthogonal ideas' – ideas of things at right-angles to each other – because he is simultaneously arguing that you're not important at all, and that you're also quite unbelievably remarkable. But ultimately this enormous universe doesn't care about you and your anxiety; it is incapable of care, abstract creativity or whistling 'Smoke on the Water'.

Floating in a tin can

For most of the history of civilization we have remained Earthbound. After a few jaunts into the air at the start of the twentieth century, the ability to fly upwards rapidly advanced and has led to a few of us flying beyond the Earth's atmosphere over the last sixty years. The human journey beyond the planet Earth has not gone far yet, but those who have travelled outside our atmosphere have had a rare perspective. Does the sense of size, insignificance and anxiety change for an astronaut? Former ISS commander Chris Hadfield first started thinking about the size of the universe as a child on cold Canadian winter nights. He was raised on a farm, where the sky was wide and artificial light did not blot out the stars. Chris used to night-ski, and the ride up on the chairlift gave him time to contemplate the stars. The fields around his parents' farm were where he started to look 'into the eternity of it'. But on this scale it was still reasonably easy for him to maintain what he describes as a psychological denial of the enormity of the universe.

After fifty years and three missions on the International Space Station, he can recognize the impossibility of the human mind wrapping itself around the huge numbers involved in measuring the size of the universe. There is a point in childhood when we become excited by counting. On the way back from school we challenge each other to announce the highest number ever counted up to, in the childish belief that we might come up with a number that no one could top, even when we realize that, with every answer, you could merely add one and it would be bigger.

'Googolplex!'

'Googolplex plus ONE!'

'Darn!'

'Googolplex times googolplex?'

'Googolplex times googolplex PLUS ONE!'

And so it would go on, until it was time for fish fingers and peas.

How big can things get, before the picture won't fit in your skull? The pebble-planet game can help to give you an illustration. Find nine pebbles (sentient or non-sentient ones will do) to represent the Sun and the eight planets. If you are short of pebbles, check that your geologist friends aren't still chewing them. With the pebbles you can model our solar system, one metre in diameter, with the Sun at the centre, and then start building other stars.

If you want to start small and include our solar system's nearest solar neighbour, Proxima Centauri, then you'll need a garden that is 140 metres long. You'll need to travel about 3,478 kilometres just to mark out the boundaries of your galaxy.

Chris doesn't believe we have the capacity to truly picture the distances of the universe. Our galaxy alone is 100,000 light years across. Even travelling in the *Starship Enterprise* at warp nine, which is nine cubed to the speed of light – 729 times the speed of light – it would still take hundreds of years to cross. The USS *Enterprise*'s five-year mission starts to look increasingly parochial. I am not sure Jim would ever get to meet any glamorous bikini-clad aliens who would ask him, 'What is this thing called love?' The most optimistic figures I can get to, for Captain Kirk crossing our galaxy, would be 137 years. That would be if he maintains a speed of 729 times the speed of light, which is a serious infraction of the laws of physics and would mean no stops *whatsoever*, even if 'Bones' or Chekov reckon there is a really interesting planet that would be worth stopping over at because the Guardians of Time live there or there's a Tribble farm.

Maybe the size of the universe brings about cosmological agoraphobia more than vertigo: to have so much around us, so much of which is empty. There may be about 200 billion stars in our galaxy and the planets that may orbit around them, but the nearest number when it comes to working out if you would collide with anything, if you set out from one end to the other and flew in a straight line, is zero. Once you are through the galaxy, there are vast voids between galaxies, though it is not utter emptiness; they are filled with matter that is predominantly made up of hot ionized hydrogen. This is known as the 'intergalactic medium'. The sky can look dense with light if you look up on a cloudless night in a place far from the illumination of the cities, but the gap between each ember is vast.

When I asked astrophysicist Jen Gupta how she perceived the size of the universe, she replied simply, 'It hurts my head' – a common reaction even amongst those who spend their days contemplating such issues. Physicist Helen Czerski thinks we attempt to create a picture in the hope of connecting to the universe, but at a certain point of zooming out into the universe, those connections become too abstract. We have not evolved to see the full picture. For many years, the furthest actual photograph taken from space was the legendary pale-blue-dot image as *Voyager* sped past Neptune, 3.75 billion miles away from Earth, back in 1990.

Ann Druyan was creative director of the Golden Record project. Both *Voyager* spacecraft have a phonograph record on them, where the music, sounds and images of the Earth are recorded – a sampler disc for any extraterrestrials who own a 1970s Sanyo sound system and may be looking for somewhere new to go on holiday.

Ann fondly remembers the day the pale-blue-dot photo was taken. It was Valentine's Day 1990. Her collaborator and husband, Carl Sagan came back to their house with a hard copy of the picture, and the two of them sat staring at it for a very, very long time. Sagan had been fighting for this image to be taken since 1981 and the official response was always 'What's the scientific value?' or 'It'll burn out the cameras.' This was going to be the very last photo of around 10,000 that *Voyager 1* had taken. What drove Sagan was his understanding that we needed to see a reflection of our true position in the universe. It would give humans some sense of perspective.

This was a dramatic sequel to *Apollo 8*'s *Earthrise* photograph of Christmas 1968. The *Earthrise* picture was enormously inspirational, but from the distance of the Moon, Earth still feels dominant. Get back to Neptune and we are, as Sagan wrote, 'a mote of dust suspended in a sunbeam'. This was the opportunity to see the whole world, the way it looks in the context of the solar system. Further back than that, out of the solar system and further into the galaxy, we vanish. Ann contributed to Sagan's meditation on the pale-blue dot. What she loves about it is that it is that moment of intersection between science and civilization – of a spiritual awareness of who we really are. She sees it as showing our true circumstances, which you don't need an advanced degree to understand. 'It's an indictment of the nationalist and the chauvinist, the polluter though fossil fuels distributor. It's all right there in that tiny, tiny one-pixel Earth. You know, I have found so many reasons to be proud of Carl, but that's certainly up there in the top ten.' The light in her eyes and the lilt in her voice, as she talks of it, show that thirty years on from that Valentine's Day all the enchantment of that image is still as strong now as it was then.

Twenty-seven years on from that image, what is now the fastest vehicle in space has taken a photo at a further distance from Earth, but this time looking away from Earth and deep into space. *New Horizons* has a speed of just over ten miles per second, travelling 683,500 miles every day. In 2017, when it was 3.79 billion miles away from Earth, it took a photo of the Kuiper belt, which contains comets, dwarf planets such as Pluto, rocks and ice – all still parochial in terms of our galaxy, let alone the universe.

An incomprehensible vastness is preferable to an easily digestible diminutive universe. Isn't it better to look out at an expanse with no visible limit than to feel the walls closing in?

However, thanks to scientific endeavour, we know there have been times in the universe's history when we could have fitted the whole universe in a single frame, and remarkably that snapshot has already been taken.

Spaghetti becomes a verb

I have become increasingly fond of cosmic microwave background radiation. Where once I had a poster of David Lynch's *Eraserhead* on my wall, now I have a picture of CMB. Both have confounded me. The CMB is an image of something whose existence I find hard to comprehend. How are we able to capture a picture of something from 13.7 billion years ago? The highest precision image of CMB was taken in 2013 by the European Space Agency's Planck telescope, which looked back into deep time and captured the microwave radiation invisible to the naked eye.

The best picture yet of what will become everything.

This radiation, the oldest in the universe, provides us with a snapshot of the universe when it was only 380,000 years old and 100 times smaller than it is now. My attachment to the image is knowing that you are in it and I am in it, and everything you have ever seen is in it – or at least the early stages of what will become us. It is a very complete family picture of the universe, its first embryo scan, when the details of what it would become may not all be there, but where everything that is needed to make a full and healthy universe is predicted, just by recording the small fluctuations in heat that we can pick up now.

I mention the peculiar sense of connection I get from this picture to theoretical cosmologist Janna Levin, and she talks about a similar excitement and connection that she feels about the existence of black holes. Her excitement is undimmed by years of academic study. 'It's amazing, I mean, some quantum fluctuation – some random quantum fluctuation – made this galaxy and we're in orbit around a supermassive black hole. It's crazy. It's bananas. It's weird because when people realized we orbited the Sun, it changed culture for ever. Somehow, realizing we're orbiting a supermassive black hole at the centre of our galaxy that was created in a quantum fluctuation is not rocking people's world.'

A quantum fluctuation is the temporary random change in the amount of energy at a point in space. It is one of the results of Heisenberg's uncertainty principle. It can seem so absurd that activity upon such a small scale can create things of such magnitude and power, but that's the universe for you, it seems. This is why reading about cosmology is worthwhile, as it can populate our minds with what is invisible to us. Janna describes black holes as being 'unlike anything else that we would ordinarily call an object in the universe. It's not a star. It's not made of, technically, stuff. It's actually empty space. And that region, the event horizon, beyond which not even light can escape, is nothing. It's like stepping into the shadow of a tree...'

I think it *would* rock people's worlds more if they knew what a black hole is, how central black holes are to most galaxies and how awesome they are in their power. My main knowledge of black holes in my schooldays was from the Disney movie *The Black Hole* with Ernest Borgnine, which, it turns out, was as useful a tool for any budding cosmologist trying to understand collapsed stars as *Steamboat Willie* would be for a natural historian studying the behaviour of field mice.

I tried to explain black holes to a former leader of the Labour Party once. I talked of Earth's gravity, a benevolent gravitational pull for creatures like us – just the right amount to keep us in place, without putting us under too much pressure. Then I explained that the more dense the matter, the stronger the gravitational pull. A teaspoonful of neutron star would weigh about ten trillion kilograms. By this time I could see that he was troubled. Politicians are so busy dealing with bureaucracy and subcommittees that they don't have enough time for astronomy, which is a great pity as a sense of the might of the universe, our fragility and the beauty beyond us could lead to better long-term decisions for the benefit of us all.

When a star twenty times bigger than our Sun, or larger, reaches the end of its life and collapses, its gravity becomes so great that nothing can escape from it – not even light. Once you have passed over that particular event horizon, that's it: there's no winching you out, at least not with the limitations of current winching technology. It has become a singularity. Briefly I threw into our conversation that our universe emerged from a singularity, and I think it was around 'infinite density and no mass' that I could see I was beginning to overstay my welcome, and my plausibility. It appeared I had delivered a new anxiety to this politician, but I think he enjoyed this cosmic anxiety more than the anxiety that comes with relentless character assassination by the mass media in the build-up to a general election.

Black holes are the reason that spaghetti needed to be transformed into a verb. Sometimes I contemplate being spaghettified. Spaghettification occurs because the increase in the force of gravity is so great, and so rapidly changing, that even over the distance of five feet nine inches, your feet will experience far greater pull than your head, and this leads to rapid thinning and stretching. If you ask the astronomically minded where they would like to die, many will choose a black hole, and for good reason. An event-horizon death-wish offers peculiar experiences and observations, both for the victim and for the observer of their demise.

Some have even postulated that falling into a black hole would create two realities for the victim – you would be instantly incinerated, but you would also continue unharmed. There is the possibility that, as you are instantaneously stretched to death at a speed that would arouse any Spanish Inquisitor, your friend who shoved you in would see nothing happening to you at all. You are almost frozen in time. Some of these concepts are size-dependent – for instance, the spaghettification is negligible in a really, really supermassive black hole.

These are not the main reason for cosmologists' hopes of death by singularity – it's the view. Time inside a black hole moves at such a different rate from outside a black hole that as you fall, you will see the rapid evolution of the universe, a sped-up spectacular of cosmic events. It would look like time was going very quickly in the rest of the galaxy. Civilizations would come and go, but it would be so concentrated, with so many colours, so much information, that it would probably appear to be bright white light. Janna Levin compares it to 'the light at the end of the tunnel in a near-death experience, but it's a total-death experience'. Space and time switch places in the black hole. There is an inevitability that we would be crushed to death by the spacetime; our bits would be torn apart, we would be brought down to our fundamental particles, and then we don't really know what happens. As ends go, it has high production values. Janna also mentioned that in this destruction our quantum particles might be blown out into a new universe as bits of information, and we would get to start the whole ecosystem over again, or we could become seeds of a new universe – the Creator.

Cooing at the baby universe

But before we develop a deity complex, let's return to the problem of size. One of the questions in Janna Levin's book *How the Universe Got Its Spots* is: Is the universe infinite or is it just really big?

What is more troubling? A very, very, very big universe – 92.6 billion light years across – or a universe that is infinite? An infinite universe means that we could have the comfort of knowing there is something familiar in the far distance, as somewhere way off there is another you and another me and everything else, because the limit of configurations of atoms leads to exactly what you see now existing across the infinite universe and, I presume, exactly the same thought patterns, so everything you are thinking now is also being thought somewhere else. Is that a comfort? According to the cosmologist Max Tegmark, repetitions will pop up every $10^{10^{115}}$ metres. This is one of those numbers that is followed by the popular statement of numerical enormity, 'That's more zeroes than there are atoms in the observable universe.'[1]

In *How the Universe Got Its Spots* Janna wrote:

> I know, of course, that the universe is bigger than my work-shop, since I can travel around London without ending up where I started. I know the universe is bigger than London, since the train I take to Manchester gets me there and doesn't deposit me back in London, barring crashes, cancel-lations and other recent fiascos. The universe is bigger than the Earth, since I can get into a plane, fly in a reasonably straight line and get out in Paris. But if I flew in a rocket out into space in a straight line, never turning or stopping, would I go for ever or would I follow the wrapping of space, to see the galaxy I left receding behind me approach in front of me?

Seventeen years since that publication, unsurprisingly she is still dealing with infinity. Two things Janna tells her students are: 'Don't just accept my answers' and 'It is right that cosmological ideas are uncomfortable.' Infinity is not something to be comfortable with; if it stops bruising your mind, then you may have forgotten what it is and what it entails. It is easy to summarize infinity in a symbol. Here it is: ∞. But once I start contemplating that dinky little symbol, I realize it is a summary of something so colossal that it seems 'colossal' is a word that is far from fit for purpose. It is something beyond size. It is immeasurable. If I think on this for too long, I fear I will fall into the toilet bowl again.

Some people believe that the universe is instantaneously infinite. They have explained that it is completely consistent with the mathematics, and that we now need to move on from that. But for Janna a mathematical comprehension doesn't stop the need to keep on asking questions. You can understand it from the mathematical perspective and see that this is the outcome that is most likely to be true, but that doesn't necessarily mean that it feels right.

She believes that what made Einstein so amazing was that he could be uncomfortable and questioning all the time. When everyone else was ready to move on, he wasn't. Understanding the size of the universe may well be a lifelong process. 'You make discoveries along the way that are concrete and unambiguous, that are mathematical, but they're usually idealized or modified or simplified. And in the end, I still don't know whether the universe is infinite or not. And I still don't know if it's a sensible construct. So I think you have to be, as a scientist, comfortable with the discomfort of not knowing.'

I find some salvation in this. I am at far earlier stage of not understanding, and I can see in Janna that years of study may progress to merely a better level of not understanding. The reasons for your not understanding infinity are clearer, your confusion is more erudite, but rather than having a blinding flash of comprehension, your knowledge may simply be like the glow of fireflies flitting around. My advice is to not be scared to approach the concept of infinity, and don't be ashamed of your struggle – the struggle is more fulfilling than passing it by. You can't really fit infinity in your head, but you can squeeze in a few of the thought experiments and mind-games about it.

To the Zooniverse: Please feed the astronomers

Chris Lintott's problem with the size of the universe is about time rather than space. Chris is an astrophysicist at Oxford University, co-founder of Galaxy Zoo and the Zooniverse. The Zooniverse is 'the world's largest and most popular platform for people-powered research'. It is a highly effective way of getting non-scientists to connect with the study of the universe and feel they can be of use, without needing a PhD. A vast gathering of volunteers may be asked to examine images of animal habitats, or far-away galaxies, and answer questions about them. It uses a congregation of human minds to look at images and to ask questions – something that computers cannot do.

Chris's interest in the night sky started in his back yard, looking at the stars and contemplating how big it all was. He had learnt about, and was fascinated by, Olbers's paradox, which is the idea that if the stars were evenly distributed in an eternal universe, there would be no dark at night, just light. We have day and night because the universe has a beginning and it is expanding.

I have found astrophysicists to be quite well balanced when dealing with the huge numbers of space – almost blasé at times. Chris thinks this may be a selection effect, in that if the vastness of space brings on a panic attack, cosmology may not be for you.

He believes that theoretical cosmologists don't have any proper understanding of how big the universe is.

In public lectures, Chris is wary of putting a vast number on the screen behind him and boggling the audience, because he doesn't think that gives them a deeper understanding of the universe. Also he doesn't believe that astrophysicists have any intuitive understanding of what it means to state that there are 100 billion galaxies with 100 billion stars in them. Astrophysicists may give the impression they have some deep and slightly mystical sense of what those numbers mean, but Chris reckons they are actually in over their heads, and they have simply got used to saying the numbers out loud. Hearing '100 billion galaxies' no longer takes them by surprise, but Chris can't really visualize that, and he doubts anyone else can.

He believes that to gain a full understanding of the universe, you don't need to think on large scales by astronomical standards. He stresses that astronomy is really simple. If he wants to understand a galaxy that is probably a 13.8 billion-year-old system of 100 billion stars, in which stars are being born and dying all the time, there is sophisticated physics to provide some answers, but there are other just as relevant questions, such as, 'What colour is this thing?' or 'What shape is it?' If you want a personal relationship with the universe, you don't have to have a deep understanding of relativity; these simpler questions create the connection, too. These are the sorts of questions that Galaxy Zoo answers – a crowdsourced astronomy project that connects non-professionals to astronomical research.

When Galaxy Zoo began, 8 per cent of participants said their major motivation for taking part in a project that involved clicking on pictures of galaxies consisted of wanting to spend time contemplating the vastness of the universe. Chris sees two specific reactions when people join the research: there's the universe's size; and there is also our position and significance – or otherwise – in it. The anxiety from these discoveries comes more often concerning the latter. People immediately jump to a position of: 'Oh no, we're on a tiny rock in the middle of nowhere.'

Perhaps we shouldn't worry about the size of the universe today, but save it for a few weeks' time, when it will be even bigger and there will be another four billion stars in it.

The comfort of the incomprehensible

There is comfort in the contemplation of vastness. Perhaps it can be a form of therapy? Neuroscientist Heather Berlin sees the moments when we become lost in the vastness of the universe as being useful to our well-being. She believes that the big numbers, and the attempts to grasp the big picture, force us to let go of our ego and let go of ourselves. We release the prefrontal cortex and lose our sense of self, which is very therapeutic. It is a moment when you may feel as if you are one with everything. Heather believes that these moments of connectedness should help all our little individual problems dissipate. 'I think it is that feeling of something more than yourself and it can lead to positive emotions and people.'

In his essay 'The Objects of Celestial Desire', philosopher David Fideler writes of 'the therapeutic benefits of observational astronomy' – something he considers even more important now than it was for the ancients, as they 'suffered far less estrangements from the beauties and wonders of the natural world'.[2] For him, astronomy is the most philosophical and contemplative of the physical sciences, and it can be 'as much an imaginative experience as it is a literal act of viewing objects in the night sky'.

One of my favourite books is the arcane *Starlust* – not erotic fiction about a salacious astronomer, but an anthology of obsessive thoughts about the pop icons of the 1980s, with particular emphasis on Barry Manilow and David Bowie. One of the more innocent fantasies concerns an intense but sweet-natured Michael Jackson fan, who writes of looking at the Moon each night and hoping that Michael Jackson might be looking at it at the same time, and his feeling therefore of being connected to his idol.

During the pandemic I know of many detached people who have used the night sky to try to briefly vanquish the sense of distance between themselves and their absent loved ones. Some people have been quite specific in this, with date-night becoming astronomy night. They arrange a time and choose a celestial object, and both participants look towards it with the naked eye or a telescope, comforted by the knowledge that their significant other is doing the same. The vast distances between us and the Moon and stars have shrunk the distance between them. The size of the universe is of little concern, when we so often have to battle with the distances between each other.

Contemplating things that we will never be able to reach can lead to us contemplating those who are out of reach or who we have lost or forgotten. That the universe should be so big and empty, and yet with at least one planet that is so rich in variety and full of questions, may make us think how fortunate we are. It is cold out there, but here there is warmth and a world full of possibilities.

Two things are infinite: the universe and human stupidity; and I'm not sure about the universe.

ALBERT EINSTEIN

CHAPTER 5

Escape Velocity – On Looking Back at the Planet from a Height

I thought of my wife and five children on that little planet.
The same forces that determined their fates worked on
the other three and a half billion inhabitants. From our tiny
capsule, it seemed as if the whole Earth was smaller even
than the space the three of us inhabited.

Bill Anders

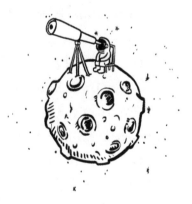

I t is good to get some perspective, and the few humans who have been into space have had the opportunity to get more perspective than most.

I have always wanted to be an astronaut. The only things in the way of pursuing such an ambition have been an uneven temperament, a fear of small spaces, a fear of heights, lack of dexterity, my total lack of any necessary qualifications, frequent fits of existential anxiety, general non-specific anxiety and a deep existential worry about ever being too far from an effective flushing toilet. I would love to see the Earth from space. I get space envy when I see astronauts floating in the International Space Station cupola window, but I realize that, like B. A. Baracus in *The A-Team*, it would require some drugged milk to get me into the capsule, and apparently NASA's ethics committee still frowns on such methods. Plus, the milk often comes back up during the dizzying docking procedure.

New knowledge changes our perspective, and sometimes this means it can change our possibilities. In the latter half of the twentieth century the human journey into space gave us new visions of what our aspirations as a species could be. This is not without its critics and its problems. There is the increasing proliferation of space junk that we leave floating behind us; fears of how the influx of private capital and private enterprise, with the race to space seemingly spurred on by the egos of the super-rich, may affect the positivity of our beyond-Earth ambitions; and the feeling that before we make big plans to populate other planets, we should really learn how to look after this one.

Perhaps we need to consider, though, that the way we should be treating this planet is powerfully communicated by looking at our world from the perspective of space. Rather than space travel being a resignation from our home, this achievement can highlight how a creature of such ambition can find inspiration to tackle its terrestrial problems from the enormity of its extraterrestrial ambitions.

In Arthur C. Clarke's story 'The Sentinel', the inspiration for *2001: A Space Odyssey*, a black monolith has been placed on the Moon by an extraterrestrial species. Should the creatures on the planet have both the wherewithal and the curiosity to explore beyond their planet and make contact with the monolith, a signal will be sent out telling the extraterrestrials that the planet is worth a visit. There is meaning here as evidence of our capabilities and intelligence.

For the first three years of my life human beings were travelling to the Moon at regular intervals. Then the journeys stopped. We returned to our default mode. No human being has stood on an organic object beyond the planet Earth since 1972, though many more have been into space. Look up at the sky and you might see the tiny dot of light that is the International Space Station carving an arc across the sky. As I write this, there are seven men and women living there, travelling around the Earth sixteen times a day.

Curiosity and conflict propelled the race to the Moon, but the success of *Apollo 11* created a sense of international fraternity. It was not only America that had gone to the Moon – *we* had gone to the Moon. The Cold War, USA vs the USSR tribal ego may have been a necessary impetus for the achievement, but it offered a viewpoint that made our border spats seem preposterous and infantile. Similarly, when the spacecraft *Perseverance* landed on Mars in February 2021, social-media posts across the world talked of 'our' achievement and what 'we' had done. Despite many of us contributing very little to the design of the Mars helicopter – this was a moment of collective achievement, a species achievement.

The great author of dystopias, J. G. Ballard, a jovial pessimist when it came to human destiny, talked of the disappointment of the space race. He felt that NASA left the poetry out of space. Ballard wrote, 'NASA spokesmen denied all along that the astronauts dreamt in space. It would have been fascinating to know what their dreams were – they might have told us something about the human race. NASA said, "Astronauts aren't the type of guys who dream." One astronaut denied he ever dreamed at all, even on the ground!' Ballard felt that there was a severe shortfall in the imaginings around *Apollo*, compared to how we considered the pioneers of aviation. 'The dream of flight entered everybody's mind. With the Apollo program and its Russian counterpart, there was no dream of Space Flight.'[1]

In my view, this is untrue if one considers the cosmonauts. The Russian space missions had a level of celebration and a mythic quality that I think shows the power of the dreams within the ambition. A powerful exhibition at the London Science Museum in 2016, *Cosmonauts: Birth of the Space Age*, displayed capsules, spacesuits, artwork and letters from the Russian public about their experience of the space race. There was a striking letter that a farmworker sent to President Khrushchev. She wanted him to know that she would be very willing to be propelled into space. She was aware that she would probably die, but if her death led to knowledge that would speed up our advance into space exploration, it was a worthwhile sacrifice.

Doctor and space historian Kevin Fong told me that he sees an apparent lack of imagination in the communication of *Apollo* astronauts as an issue of bandwidth. The *Apollo* astronauts were under tremendous pressure to succeed in this pioneering achievement. The intensity of the focus on the job at hand may have left little or no room for further contemplation. Kevin considers that it might be like intense medical procedures with which he has been involved, where the focus required means that you are seeing only a very small part of the picture. It is difficult to free up enough bandwidth to have any sort of quasi-philosophical experience in moments of such focus. The astronauts were also chosen with great rigour – not for their spiritual or poetic inclinations, but for their physical prowess, psychological level-headedness and ability to follow orders.

Kevin has watched astronauts up close and has also made an award-winning series, *13 Minutes to the Moon*, a remarkably comprehensive look at the preparations for *Apollo 11* and its eventual landing on the Moon. He says the one question to avoid, when interviewing *Apollo* astronauts, is 'What did it feel like to be on the Moon?' On the whole, 'feeling' questions got short shrift.

Andrew Smith, author of *Moondust*, believes that going to the Moon didn't change the *Apollo* astronauts, but it allowed them to be much more like the people they already were. Neil Armstrong was utterly focused on the technology and the engineering achievement, rather than on the enormity of the experience of being the first human to stand on another world. Kevin says that it was well known that if you sent a letter to Armstrong saying, 'I'd love to talk to you about what it was like to stand on the Moon and look back at Earth', he never replied.

If you wrote, 'I'd like to talk to you about the landing gear on the X 15', he would reply straight away. For him, the technology was more beautiful than the view.

They weren't all like Armstrong, though. *Apollo 14*'s Edgar Mitchell had a quasi-religious experience on his trip and later became fascinated with psychic research. Buzz Aldrin took Communion during the Moon landing, though that was not highly publicized at the time. Fred Hayes on *Apollo 13* looked back at the Earth and said in wonder, 'You know, guys, the world really does turn.' The *Apollo 13* Capcom Joe Kerwin replied, 'Maybe we'll call that the Hayes theory' and then everyone burst out laughing. Hayes's expression of wonder was subject to friendly mockery. Hayes said afterwards that it was almost as if in that moment he 'broke character'.

The military mindset of many on the NASA teams, and the barrack-room banter, perhaps made moments of philosophizing, contemplation or wonder something to keep to yourself. Some *Apollo* astronauts, though, seemed to possess those traits of empathy and connection more than others, such as Jim Lovell, of *Apollo 8* and *13*, and *Apollo 9*'s Rusty Schweickart. Lovell saw himself as an explorer, which is why he called the *Apollo 13* command module 'Odyssey'. It was Lovell who grabbed fellow astronaut Bill Anders on *Apollo 8* and dragged him to the window to take the *Earthrise* photograph. No matter how accomplished a pilot he was, it was Lovell who really got a sense of the bigger significance of what they were doing.

The birth process is triggered

In Andrew Chalkin's comprehensive book on the *Apollo* missions, *A Man on the Moon*, Rusty Schweickart is described as a 'red-headed, irreverent rookie'. I first met Schweickart at the Royal Albert Hall – not a regular haunt of mine, due to my failure to pass my clarinet exams. It was, however, a night of celebration of space exploration, mixing music, comedy, science, lasers and astronauts. I was anxious about my first meeting with an Apollo astronaut, but I was immediately put at ease by Rusty's playfulness. Now, when I get an idea for some strange event in which an Apollo astronaut would come in handy, I contact Rusty and this is a typical reply: 'It wouldn't be you if it wasn't weird. And, really, I like weird. Especially cosmic weird... BIG WEIRD!'

Rusty was the youngest *Apollo* astronaut. His mission, *Apollo 9*, was a pivotal mission where they tested out the lunar module and rendezvous manoeuvres. Rusty was also to attempt the first *Apollo* spacewalk. As the launch approached, Rusty's preparation for space was not merely concerned with the physical and psychological necessities for the mission itself, but also had a larger philosophical element. On solitary Sunday evenings he would listen to great works of music and read literature and philosophy as he contemplated what his *Apollo* mission meant – his human journey outside our atmosphere. He was aware that he was about to become a very active part of an historic event, not just of *Apollo 9*, but of the entire space programme. As time went on, he began to think in larger and larger terms of Earthkind, the evolution of life on Earth and the possibilities of the journey beyond it.

He showed me some of the tiny scrolls that he took into space. He selected twenty-three quotations and had them printed on parchment and bordered with gold leaf. The sources included the Bible, Julius Caesar, Elizabeth Barrett Browning and Bertrand Russell: 'Curiosity is the primary impulse out of which the whole edifice of science has grown.' Rusty wanted to take the wisdom of humanity with him into space.

On the mission itself, whilst Rusty was on a spacewalk, there was a technological hitch while preparing an engineering test of the first lunar module. For a few minutes Rusty was fortunate to have some free time to himself. He did not have a Damascene moment during his free time floating, but those few moments crept up on him over the coming years. When he explains it to me, it makes me think of Darwin's voyage on the *Beagle*, that was followed by decades of gestation, of ideas stirring and stirring in his head, until the theory of natural selection fitted into place.

For Rusty, it was a meeting with a philosopher who was cynical about the space race that would see his thoughts becoming fully focused. Before the launch of *Apollo 17*, the final manned mission to the Moon to date, which Rusty was involved with although not part of the crew, he was introduced to the philosopher William Irwin Thompson. Rusty describes Thompson's attitude to the *Apollo* missions as 'technology and bullshit' – a nationalistic enterprise with 'all-American heroes'. Over lunch, Rusty noticed Thompson's gears change. Thompson realized that Rusty had done a bit of reading, that the *Apollo* missions were not simply some John Wayne gung-ho adventure and Rusty wasn't just some unthinking air jock, so they had lunch again the next day before the *Apollo 17* launch.

Due to a delay, this was the only night-time rocket launch and it was a spectacular sight. The suspense built and, when *Apollo* finally took off, the exhaust plume looked exceptionally bright in the darkness. The noise made by the *Saturn V* rocket is 204 decibels. It rattled windows for miles around. It can seem as if the rocket itself could not survive the noise of such a blast. It is the loudest man-made sound save for the Tsar Bomba, the most powerful nuclear bomb ever created. The sound energy of the launch is distributed through frequencies from low to very high. 'The longer the string, the lower the frequency,' as Rusty explains. 'Some of the frequencies are below the audible, so what you hear is deafening, but you really feel it too, your internal organs oscillating at frequencies below audible.'

The day after the *Apollo 17* launch it was clear to Rusty that Thompson's cynicism had been eviscerated by such a profound personal experience. Rusty tells me, 'It just absolutely blew all his circuits.' Now Thompson saw Rusty as 'a fellow traveller' and invited him to speak at the yearly gathering at his commune, the Lindisfarne Association on Long Island. It was decided that Rusty would talk about the implications for humanity of space travel – of the philosophy and psychology of what it meant to humanity to be able to look back on itself from space.

As the months passed, Rusty kept thinking that he really must write that speech, but procrastination usurped his good intentions. The afternoon before his speech, he went up on a hillside to make notes, but he heard someone playing a drum so beautifully that he got lost in the rhythm and reverie. By the time he walked onstage the next day, he had one card with about four lines on it. He had no idea what he was going to say. Then he started to talk. 'I suppose the reason that a nice, down-to-earth astronaut like me is here, in a far-out group like this, is somehow to share an experience which man has now had.'

He quoted the poet Archibald MacLeish, who wrote of the *Apollo 8* mission in *The New York Times* that now we saw the Earth as it truly was, 'small and blue and beautiful in that eternal silence where it floats'.[2] He told the audience of those moments when you could look out of the window:

> You reach down into the cabinet alongside the seat and you pull out a world map and play tour guide. You set up the little overlay which has your orbit traces on it on top of the map, and you look ahead to where you're going, what countries you're going to pass over, what sights you're going to see... Every hour and a half you go around the Earth and you look down at it. And finally, after ten days, 151 times around the world, 151 sunrises and sunsets, you turn around and you light the main engine again for the last time, and you slow down just enough to graze that womb of the Earth, the atmosphere.[3]

By the end of the talk, half the audience, as well as Rusty, were in tears. This was the end of his five-year gestation period – all that thinking that he had experienced as he floated outside the capsule, those few moments given to him by a technical glitch, came together as a coherent revelation. He saw the speech as coming from his subconscious, lured out by the connection he felt with the people in the room. From that point on, he saw himself as the astronaut who could connect with the hippie-commune era. Rusty believed it was important that the counterculture didn't get left behind, that it didn't see technology solely as the concern of the political right, of nationalists; that the advance in man's capabilities was not seen as the tool of oppression or as ideologies that were in conflict with counterculture aims.

He thought more thoroughly about the metaphor of birth: the idea that humanity was beginning to take its first tentative steps away from its mother, from the Earth. A 'fetus' is demanding in terms of resources and energy, and producing all kinds of waste that the other has to process, and as it reaches its limit, the dangerous birth process is triggered. 'Your potential can only be attained by leaving the mother, going beyond the limits of what it can provide in this little cocoon called my womb. And that's what's going on with the Earth.' He began seeing that parallel more and more often. 'It is a metaphor that you can push pretty damn far, frankly, and it still retains legitimacy. And that is, in fact, what's going on, and that the Earth is, in fact, our mother. And we talk about Mother Earth, we do it in some dumb dipshit way; in fact, it's true,' he says.

Rusty is drawn to the ideas of Big History – the history that David Christian espouses, which does not simply zoom in on the actions of humans, but puts humanity in the whole flow of the universe from the Big Bang and beyond.

In the year of the *Apollo 11* anniversary, Rusty knew that, as the youngest *Apollo* astronaut, he would be invited to speak at many events and he really didn't want to keep talking about *Apollo 9*. It was a wonderful experience, but he knew the world was a lot bigger than his mission. If he was going to focus on any

Apollo mission, he was going to be inspired by *Apollo 8*, that first observation of the Earth coming up over the horizon – an entirely unexpected sight on a mission that had been rigorously planned. The crew had been orbiting the Moon for six hours and then they looked at the forward horizon instead of the receding horizon, and up over that horizon came the Earth. As Rusty puts it, 'They're the first people outside the birth canal looking back and seeing Mom and realizing the process that they're part of.'

In terms of Big History – where physics generates chemistry, and chemistry allows evolution into biology – Rusty defines that biology as something 'that gives a damn'. It wants to survive and it wants to reproduce, and we are a manifestation of this huge process. Rusty sees our part in this as one with huge responsibility, a question of whether we will join a community of life in space? His philosophy on our responsibility on Earth, and beyond, is that we are not the passengers, but that we are the crew. We're not just sitting on this thing and being carried along, having a nice time; we are responsible for maintaining and driving it.

'Whitey on the Moon'

A frequent critical question about space exploration is: Is it worth it? When the world we inhabit is so full of suffering and seemingly intractable problems, is it worth spending billions on putting men and machines into space?

I can find myself conflicted about human space ambitions. I love to watch a launch, and can even get excited imagining a launch. I was at Cocoa Beach in Florida on the 50th anniversary of the launch of the *Apollo 11*. At the exact time of the anniversary of *Apollo 11* launch I was sitting on the beach, which, with my pale skin and wilting demeanour, was no mean feat. The last time previously that I had sat on a beach in Florida was 1988. I stayed under the shade of a tree, but for the next week, my nose, ears and cheeks scabbed and peeled as if I were about to burst out as a butterfly. Sadly, all that was revealed was a slightly more freckly me and I still needed an air ticket to get home, but it was worth risking going into the sun for a second time. So I looked to where the *Saturn V* rocket would have been. I could feel the ghosts of the crowd that would have filled the beach that day, and a memory of the gathered awe and suspense felt palpable. Like my moment of touching the Standing Stones of Callanish, memory and imagination flooded my senses with a vivid collage of a launch that would send three men hurtling towards the Moon and confirm a new human possibility.*

I was lost in this reverie of exploration when my phone beeped and I saw that someone had sent me a link to Gil Scott Heron's song, 'Whitey on the Moon', in which he expresses anger that while white men are on the Moon, black people in rat-infested housing can't even afford healthcare. Prince expressed similar sentiments in 'Sign o' the Times'. That's the problem with human ambition; it can leave people behind. As

---

* Actually I was a bit jet-lagged and got the time difference totally wrong, and so all that psycho-geographical event that was going on in my mind was a few hours too early. What I should actually have been experiencing was some hot-dog sellers folding out their stands in preparation for the rush.

William Gibson said, 'the future is already here, it's just not very evenly distributed'.[4]

Our grand space ambitions can offer a beautiful tapestry that may be draped over other human problems, but you can also wonder if it is all sleight of hand. Would the money spent on rocket fuel ever have found its way to the impoverished and marginalized? Or does demonstrating such spectacular technological expertise inspire humanity to reach further and, by doing so, also strengthen the focus on the departure point as well as the destination? My belief is that the cancellation of human adventure and exploration will not lead to the money saved going into welfare programmes and pulling people out of poverty and despair (there is also the separate issue that there is a financial return on these missions that often outweighs the cost). Plans of voyages to Mars are not what is getting in the way of building a better and more equal society. Inequality is predominantly a separate issue, but I would hope that those private companies that are now financing space missions have not built up the fortunes needed to become extraterrestrial by skimping on tax or employee rights and benefits.

Our journeys into space should give us a perspective about wanting to build a better world. At least from the perspective of the astronauts I have met, their privileged view of the planet has sharpened their ambition to deal with the problems of human civilization. The sharpened minds of those who are attempting improbable feats of exploration may also discover ways of directing these innovations towards the most pressing issues on our planet, rather than merely working out an escape route.

Political philosopher Hannah Arendt worried about our entry into space. When the first artificial satellite Sputnik 1 was launched in 1957, she published a plea asking us to stop and think about what it really meant to go beyond our atmosphere. She was worried that sending humans up into space – actually touching the night sky for the first time and being able to look down on the Earth in its entirety – would give us an ultimate Archimedean viewpoint of our planet. She saw this as a point of view removed from the system that humans were inhabiting; that the scientists would make us completely cold and detached from ourselves, looking back on the planet as if we were a slide under a microscope, thereby diminishing our humanity and our empathy for the Earth and the people on it. For the astronomer Stuart Clark it is 'one of the sort of delicious ironies of history, the complete reverse happened; the most important thing that came out of the early spaceflight – and, I think, probably continues to come out of spaceflight – is our view of the Earth and our understanding of Earth as a fragile, integrated whole that we have to steward and take care of.'

'Spaceship Earth'

The overview effect for the International Space Station astronauts is very different from that of being on the Moon. Firstly, the ISS is far closer to the Earth. Astronauts can sit in the cupola and watch Earth go by. Secondly, there is time. It is not as if the ISS is about lounging around in casual spacewear, but many are up there for months on end. Unlike an *Apollo* mission, this is not a journey so much as a change of location and possibilities.

The view below can become familiar, with each stretch of sea or range of mountains instantly recognizable. I think J. G. Ballard would have been more impressed by ISS engineer Nicole Stott than by the *Apollo* astronauts, who underwhelmed him. Ballard believed they were devoid of imagination and dreams (something I do not believe is true). Nicole describes herself as an artist, astronaut, Earthling. She has spent 103 days in space and was the final astronaut to return to Earth on the Shuttle, and was part of the first NASA extraterrestrial tweetup.

She reckons that she was no great shakes at geography when she was at school, but that has all changed. The crew would challenge each other to name deserts as they travelled above them, or be the first to spot the Amazon River or the Nile Delta. There was always an extra kick in seeing Florida from space, because that is the state Nicole considers home. It is a reminder of the inspiring phrase popularized by the architect and inventor R. Buckminster Fuller, who wanted to remind us that we travel on 'Spaceship Earth'. We are not simply hanging still in a fixed position in space, but travelling through space. Everything is in motion, from the atoms in the seat underneath you to the galaxy that you are in. Everyone on this planet is upside-down and everyone is the right way up, all at the same time. Thoughts like this don't give me cosmological vertigo, but I do get a bit of galactic teetering – a brief imagining of sloping off the Earth at an angle, as gravity fails me and the atmosphere thins.

Nicole sees the operating systems of the ISS as a reminder of the natural operating systems of the living Earth and the delicacy of that balance. 'We build these mechanical systems in space, where we are essentially doing our best to mimic what Earth does for us. Every day up there, we're acutely aware of how much carbon dioxide is in our atmosphere. How much clean drinking water we have. What's the integrity of our thin metal hull? All of those things that are a perfect analogue to how we should be behaving down here on Earth. I love the thought of that hull. That's the only thin metal that's between us and the deadly vacuum of space, and the same thing with our thin blue atmosphere.'

I feel that for many astronauts there must be a sense of loss when they return to Earth, especially if they know that it has been their final time. Wouldn't you sometimes wake up in your terrestrial bed believing you could just float up out of it? Alone in the park, wouldn't you sometimes make a little jump in the air and wonder if you might stay suspended above the grass for a few seconds?

When Nicole watched her first electrical storm from space through the cupola window and tracked the movement of the lightning strikes, she felt the impact of realizing that there was not a single thing that was not connected. She remembered the Florida storms of her childhood that had drenched or illuminated her. Now she looked down at the Earth at night and watched these strikes across the planet. They looked like neurons firing across a brain. She could see a path from where she was looking to the other side of the planet, wrapping around it.

Back on Earth, Nicole looks up and thinks, 'Wow, the atmosphere looks like it goes on for ever, when in reality it's like a veil and less than the thickness of an orange peel on the planet as a whole.' She wants everyone to find a way to feel that. 'It can be a positive influence on the decisions that you make about how you want to live your own life and share it with the other lives that are here on Earth.'

When she returned from space, Nicole felt people almost willing her to feel melancholy and moribund. 'Oh, so that's space over – the pinnacle of your life is behind you.' But she wouldn't slouch into retreat. She feels that, given what she's experienced, she wants to do something even better. Now she works at finding ways of mixing art with ideas of space exploration to connect children to the story.

'That's home.'

When Nicole was visiting a paediatric cancer centre, she struck up a conversation with a nine-year-old patient. As they were painting together, the girl started talking to Nicole about what she was going through in the hospital with her treatment, and how similar that must be to Nicole in space. 'I come into the hospital and I have to stay here for a long time. And I don't get to see my mommy and daddy and friends the same way, and I get to eat all different kinds of food, and my body's changing and they're doing all different kinds of tests on me. And I think you guys have this radiation thing in space, too, that we have to deal with here.' This girl had found all these parallels to her cancer treatment and being an astronaut.

'We each personally have a talent that can help make life better for everyone around us. And that's why I loved going and living and working on the Space Station. I wouldn't have launched on a vehicle with seven million pounds of exploding rocket stuff underneath me, with a seven-year-old son at home, if I didn't think there was greater good to come from it. And I think that's really, in the end, what it's all about. How do we find something in ourselves to help make life better for everyone around us?'

'Why would they choose a girl who works in a chocolate factory and has got brown hair?'

Helen Sharman had a similar reaction to her own journey. Helen was the first Briton in space, twenty-four years before the next one, Tim Peake. When Peake went into space, Helen's achievement seemed to slip from the minds of many people in Britain, in much the same way as, when a British man wins the Wimbledon tennis tournament, Virginia Wade's victory in 1977 can slip the mind of the casual sports historian. Tim Peake, to his credit, would frequently correct this error (the one about Helen Sharman, not about Virginia Wade. I am sure he would correct the one about Wade too, it's just that he doesn't get asked about tennis so much).

Helen's parents mixed pragmatism and theory; her mother was a nurse and her father worked in academia. He was the sort of father who really would explain why the sky was blue. Helen would look at a leaf and say, 'I find it really interesting that I am looking at this leaf, but that isn't really the leaf now, that is the leaf a fraction of a second ago', and then her mother would tell her to stop being so ridiculous.

Unlike many astronauts, Helen's journey into space was not the result of a military career, or even a driving childhood ambition. Instead, listening to the radio one day as she drove home from her work for Mars Confectionery, she heard an advert on the radio: 'Astronaut wanted, no experience necessary.' Had she been driving under a bridge at that point, she might still be perfecting the chemistry of caramel; 13,000 people applied and Helen was the successful candidate.

Her advice, if you become an astronaut, is to 'look out of the window as much as you can'.[5] When she looked out at the Earth, she wondered why we find it so attractive. Is it because we evolved there? Is it because it is our home – where everything we have ever done, and everyone we have ever known, is? Why is blue so attractive to us? What is the evolutionary advantage of it all?

She can get frustrated, reading science that is incorrect about the physics of space travel, such as the children's book which explained that humans float in space because there is no gravity.

Gravity doesn't stop, once you leave Earth's atmosphere; it just gets weaker, the further away from Earth you go. Helen likes to explain how relatively close astronauts on the ISS are to the Earth, and therefore how strong gravity still is, and at this point she can see an audience thinking, 'Well, this doesn't make sense.' Then she explains about falling around the Earth. She might have felt weightless, but gravity was actually still acting upon her and pulling her towards the Earth. Astronauts don't feel that pull, because they are in free fall. Helen sees understanding as a way of overcoming anxiety and fear, of seeing the world as it is. It enables people to ask the right questions and make more informed decisions. She uses her experience in space – her stories of how the body changes, for instance – to tool people with critical thinking and to ask 'Why can that be?' and then work through to an answer.

The dominant transformation in Helen after her mission was that it made her focus on the importance of people, rather than of things. Before her training trips to Russia in the 1980s she felt very much part of the rat race, striving for material gain. She was well paid by Mars Confectionery, had a sporty car and a hunger for big motorbikes. Russia was very different. Her friends there didn't have cars and, at the training centre in Star City, the community spirit was strong. The support of the families of other astronauts, the doctors in the community – everybody supported each other and came together. And then, once in space, nothing mattered other than people.

Like Nicole, Helen now thinks, 'What can I do to make best use of my experiences in space?' She uses her platform to promote the importance of thinking logically; of not being misled by scaremongering headlines; of finding the right ways to debate ideas, because, she says, for a democracy to be a democracy, it is vital to have access to information and the ability to understand what it means.

Are astronauts immune to existential anxiety?

When nine-year-old Chris Hadfield saw Neil Armstrong on the Moon, he realized that was what he wanted to do, too. When he was thirty-six years old, he made it into space. He has spent a total of 166 days in space, on three ISS missions. What has always impressed me about astronauts is their incredible calm. Obviously being a hothead or suffering from anxiety attacks whilst living in a tiny capsule with six other people is not recommended. But I wondered whether Chris ever experienced existential anxiety as an astronaut, or whether it was as unlikely for him as it seems to be for many particle physicists?

He thinks that, in order to have existential anxiety, you have to make it personal. If you are secure here in the dimensions of the surface of the world, then you generally don't allow the idea of the centre of the Earth to bother you. We all know that we are standing on a very thin crust and, not very far down, there is liquid rock for thousands and thousands of kilometres, but unless we stand inside a volcano, it doesn't affect our lives, so we can comfortably ignore it. If we see the lava flowing towards us, then it becomes personal. Even inside a spaceship, Chris says, you are psychologically in a cocoon.

The nearest he came to a sort of personal interaction with existential anxiety was when he went outside the capsule on a spacewalk. Suddenly there was so little artifice between him and eternity – just a cloth suit and the Perspex of his visor. He could really picture where he was, because now the Earth was not Mother Earth with him sitting in her lap, but a planet in the distance: a close distance, but still not where he was. Psychologically he was split. He was no longer in his ship and he wasn't standing on Earth, he was out there in eternity.

Yet even that didn't overwhelm Chris. The dominant feeling he experienced was 'of being infinitesimal, of being so incredibly tiny in comparison to everything else, of being a speck, both physically and temporarily'. He felt minuscule, but also enormous. Minuscule not merely due to the size of the galaxy he was in, but also thinking of the time – his possible 100 years of existence, versus the 4.5 billion of our planet or the fourteen billion of the universe. The enormity came from him thinking about human cleverness, the ability to invent and create that had put him here, floating in space. Because of human ingenuity and the ability to imagine, Chris was now starting to experience personally something that was going to alter his perspective for ever.

J. G. Ballard did not have the same adventurous and ambitious hopes as astronauts like Rusty Schweickart and Nicole Stott. For him, our post-terrestrial dreams would remain dreams, and occasionally could become nightmares. He said, 'It may be possible that the human central nervous system doesn't have a designed capacity to explore Outer Space. Zero gravity perhaps recapitulates on the unconscious level various archaic fears in the human mind – falling off the branch into the jaws of the predator below.'[6]

Claude Nicollier, whose achievements as an astronaut include an eight-hour spacewalk while servicing the Hubble Space Telescope, sees human travel to space as part of a process of the Darwinian evolution of our species. He says, 'This is a step in the evolution of humans that will lead us to a situation where we have a better survival possibility long-term, pretty much like part of life left the water, but 300 million years ago, and went on land. This was, in a way, a biological necessity, and I think going to space is, in a way, a biological necessity. We can't avoid it. It happens because it gives us long-term, and a higher chance of survival to humanity. When I say long-term, I mean decades, centuries or millennia.'[7]

I hope our ambitions remain, and that our travelling into space will inspire us to focus on the issues and battles that we have on our own planet. If we can ensure clean water for our astronauts, surely we can strive to ensure clean water is available for everyone on Earth? I still love daydreaming about living in some exploratory pod on Mars or beyond, but I now also spend more time thinking about my good fortune to be living on Earth, rather than striving to be a Martian. I am enchanted whenever I see images of Mars or Jupiter or Venus, but I am also aware that Earth offers a view of incredible variety and scope for the smallish planet it is or, indeed, for any planet that we know of so far. One day we might be able to forge some sort of life on other planets, but I believe we will always be most connected to the planet of our origin.

When I asked Rusty Schweickart where he would most like to go to in the universe, he said, 'The next meeting of the Federation of Sentient Culture, somewhere in our spiral arm of the Milky Way.'

We need to make sure that we are worthy of their attention, and worthy of an invite.

And behold
The blue planet steeped in its dream
Of reality, its calculated vision shaking with the only love.

JAMES DICKEY

CHAPTER 6

Why Aren't They Here?
Or Are They...? – On Waiting
for Our Alien Saviours

Klaatu barada nikto.

Klaatu in *The Day the Earth Stood Still* (1951)

Dear Sir: The planet Venus is inhabited with beings who
look just like ourselves. No doubt you cannot see this what
with all the education you may possess. For when one
becomes educated things become complicated and that
simplicity is non-existable.

A letter to the astronomer Patrick Moore

S hould we be terrified because we are alone in the universe
or terrified because the alien invasion is only around the
corner?

John Shepherd was brought up in rural north Michigan by his grandparents. He became fascinated by the idea of communicating with extraterrestrials. His story is told in the short documentary *John Was Trying to Contact Aliens*. He built his own machinery, slowly engulfing the house with communication technology. He sent out coded signals, as well as broadcasting Afro-pop and reggae. He explained, 'My interest is in finding out the unknown, and the unknown is just that – unknown, and you search, and you continue searching, because of your desire, because you know there's something there.'

He described his lifestyle as remote, and different from those who lived around him. His body was in a quiet little local community, while his mind was travelling the cosmos. John is also gay, something that was a tough reality for him in rural Michigan. Just as he believed in the possibility of communicating with life 'out there', so he believed there was someone on Earth for him to truly communicate with. The project gave his life meaning and inspiration, and the chance to communicate ideas with like-minded people. John has not connected with aliens, but his attempts to do so have connected him to other human beings. His project is now dismantled, but in 1993 he met the right man. Both looked at each other and thought, 'Here is someone different.' They have been together ever since.

Loneliness, both Earthly and beyond, is one of the great quandaries of being human. Arthur C. Clarke wrote, 'Two possibilities exist: either we are alone in the universe or we are not. Both are equally terrifying.'[1] Are they equally terrifying? I am not so sure.

In the realm of science fiction, extraterrestrials normally either arrive from outer space with a matriarchal hand to say, 'You lot need to calm down'; or with a patriarchal fervour, sucking out our life-force and inhabiting or eating our bodies. Sometimes they haven't even bothered to say hello, they have just used our rock as a rest break. For Steven Spielberg, the enemy has been an aggressive and anxious humanity, rather than those who visit us. His aliens have been inquisitive and benevolent in *ET* and *Close Encounters of the Third Kind*. They have been both childlike and technologically advanced.

Ridley Scott gave us a non-stop killing machine with acid for blood. Douglas Adams gave us aliens who spewed ghastly poetry and destroyed all of human civilization, because we were standing in the way of progress and a shortcut through space. There are some aliens that come to save us from ourselves, such as in the original *The Day the Earth Stood Still* and *Arrival*, based on Ted Chiang's *The Story of Your Life*. If there is no God, could extraterrestrials be the alternative? Philosophers, psychologists and religious scholars have proposed that the experience of encountering what people believe to be an alien craft is the modern equivalent of experiencing an angelic apparition. Could an alien species be the parent figure that can show us the way forward – one that has been through this brutish, aggressive and evidence-denying period and come out the other side?

Stephen Hawking was less optimistic about the outcome of such an extraterrestrial union. He was wary of sending out signals across the universe that might act as beacons to intelligent species, because he feared a superior species would treat us exactly as we have treated those species that we have believed ourselves superior to. He felt that if an alien species had the technology to travel across space and visit us, they would most likely be at a point beyond our own neural evolution. 'Such advanced aliens would perhaps become nomads, looking to conquer and colonize whatever planets they could reach,' he said. 'Who knows what the limits would be?' In the 2016 documentary *Stephen Hawking's Favourite Places*, he reiterated his views: 'Meeting an advanced civilization could be like Native Americans encountering Columbus. That didn't turn out so well.'

However, what unsettles me most is not the belief that I might one day be enslaved and then farmed for alien consumption (at least becoming a snack would give my life purpose), but the thought that there is complex life beyond this planet that we will never meet or communicate with. It is not scary being alone because there is no other intelligent life; for me, it is being alone knowing that complex and intelligent life is scattered through the universe, but utterly unreachable. At the very least, it's disappointing.

In periods of peak insomnia, I decide to stride out across the galaxy. I choose a star from the inaccurate map in my head and imagine an Earth-like planet orbiting it. I see all forms of life flourishing on it. I try to stretch my imagination, escape from the boundaries of our presumptions about what is needed for life and how it might appear, and then I usually end up with something in my mind akin to a mid-1970s prog-rock cover, perhaps Yes or Uriah Heep.

Are we only a galactic lay-by?

Enrico Fermi was the creator of the world's first nuclear reactor and won the Nobel Prize in Physics in 1938. After attending the Nobel ceremony in Stockholm, he did not return to his native Italy as, under the racial laws being introduced by Mussolini and the Fascist government, being Jewish would have put him in great jeopardy in his homeland. Although he is lauded as a great physicist for his pioneering work in both quantum theory and nuclear physics, his name is most likely to come into conversation because of an extraterrestrial question. During the UFO boom of the early 1950s, Fermi wondered: if life is likely in the universe and we live in a galaxy of billions of stars with possibly hospitable planets, where is the other intelligent life, and why hasn't it made its presence known? It became known as 'Fermi's paradox' and can simply be summed up as: 'Why aren't they here?'

Part of the answer might be that 'they' are as technologically restricted as we are. We might be broadcasting, but who is to say if 'the others' are on the same bandwidth? The decade after the first UFO boom of the 1950s, some people made money not by asking the question 'Why aren't they here?', but by suggesting that there had been centuries of covering up alien visitations to planet Earth.

At the forefront of this movement was Erich von Däniken, whose books *Chariots of the Gods* and *Return to the Stars*, and multiple spin-offs and a film, piled up the evidence for alien interventions in ancient civilizations. The evidence did not often bear close scrutiny, but many didn't care about close scrutiny, as the will to believe was greater than the desire to scrutinize. Von Däniken's idea became known as the 'ancient astronaut theory', although this was not a theory like the theory of evolution or of general relativity; it was a theory like 'Bill Gates wants us all to be injected with a vaccine against Covid-19 because he has put lots of tiny robots in the vaccine and they will control our minds and eventually trap us in an Xbox reality like *Dino Crisis 3*.' In his introduction to *The Space Gods Revealed* by Ronald Story – a book that held up these claims to the sunlight of evidence-based scrutiny – Carl Sagan wrote that von Däniken's argument was 'that our ancestors were too stupid to create the most impressive surviving ancient architectural and artistic works'. There was an awkward racist presumption that those simple non-Aryan people of ancient times would not have had the geometric wherewithal to create such objects of wonder. It put a dampener on artistic imagination, too, taking a literal view of the images on the walls of sacred buildings that took the form of hybrid human/reptile/bird creatures as being representations of our human/bird/reptile/cat alien visitors. Sagan believed that von Däniken's appeal was 'theological in origin', and that in perilous times we might hope that there could be something beyond the Earth that would come and save us from ourselves. For von Däniken's theory to thrive, it required suppressing archaeological and historical knowledge that contradicted his presumptions.

I have picked up many books in flea markets and charity shops from the ancient alien-civilization boom of the 1970s, which entice you in with their back-cover blurb asking questions like: 'Did a giant spaceship abduct a British regiment in 1915?', 'Are spacemen helmets drawn on a Mayan codex?' and 'Was the cabbage artificially created so that it could directly communicate with the Sun, having been created in artificial light?'

The success of this industry shows how strong our desire is to believe in an extraterrestrial intelligence that could communicate with us. Even now, fifty years after von Däniken's frequently refuted, but highly profitable claims were published, the seaside resort of Blackpool is proposing a £300 million leisure complex that will include a major attraction based around von Däniken's books. The question is: who will build this seaside leisure complex – will it be the humans of Lancashire or will extraterrestrial intelligence strike a deal to build the burger joints and helter-skelter?

Maybe we don't need to wait for the aliens – maybe we *are* the aliens? Some astrobiologists believe that the origin of life on Earth is from life in space: a theory called panspermia. Caleb Scharf, director of Columbia University's Astrobiology Center, wonders if we may be Martians. Did Mars possess an atmosphere that was more conducive for the formation of basic life than Earth, and then debris from Mars collided with Earth and brought life here? Or, if we find remnants of life on Mars, could it have originated on Earth? The idea of the interplanetary exchange of genetic material has a beauty to it: the contagion of life due to explosive and destructive planetary events; the potential of life spreading across the universe on rocky fragments, some falling on barren, inhospitable ground, some landing on soil that has been waiting for something to stain it and spread through it. The idea also means that you only have to get life started once or twice, and pretty soon there would be life everywhere. Life itself could be a highly contagious disease.

What if the aliens know we are here and they don't care – the 'rest stop' theory of visitation? One of my favourite novels is by the great Russian science-fiction writers Boris and Arkady Strugatsky. *Roadside Picnic** is the story of an alien visitation to Earth, but the aliens take no notice of us whatsoever. The visitors pull up to have a snack, fly-tip their rubbish and carry on their way without the slightest interest in the life that may exist around them. It's like pulling over at a lay-by outside Luton, emptying your ashtray and leaving your sausage-roll wrappers, then moving on without the slightest acknowledgement of the existence of Luton and its many amenities.

For some, theories of life on other planets may fit with their belief that life is God-given, though some faiths face numerous theological conundrums, should life not be the sole preserve of Earth. If Earth stopped being God's favourite planet, with his favourite species, and the highly unlikely Venusians looked like the favoured ones, we would have a reason for the first interplanetary religious war. We can only hope that the first aliens we meet are cheery godless Buddhists, with no hankering to covet our material possessions and carrying reasonably non-invasive incense sticks.

But the flipside of imagining a universe speckled with life is to consider: what is galactic loneliness? I have felt it when I look out at the sky on a clear night. It does not feel the same as that teenage

---

* *Roadside Picnic* was the basis for Andrei Tarkovsky's masterpiece, *Stalker*. It is inspired by the book, but offers a different narrative, structured around the world created by the Strugatskys. You should also read Geoff Dyer's *Zona*, based around watching the film *Stalker*. Please finish this chapter at the very least before moving on to these, though.

loneliness* when you believe that you will never find anyone who feels the same as you – someone you can really communicate with. For me, it is fed by confusion and melancholy wonder, rather than desperate need.

It is loneliness with a sense of intrigue. When I look up on a clear night, I have feasted on the excitement of me being the first interaction with a photon of light that has possibly been travelling unimpeded for 4,000 years. I start to imagine the unseen planets that orbit around those stars. I wonder if, with a very powerful telescope, I could look directly at something 2,000 light years away, which is looking directly at me through a similarly very powerful telescope. I can start to think about living beings billions of miles away. Would I be able to understand them? Would they be able to understand me?

They came in peace, we gave them noise

The SETI Institute is a research centre devoted to the Search for Extraterrestrial Intelligence. One of its most famous outings in popular culture was in the movie *Contact*, starring Jodie Foster as a scientist who finds evidence of extraterrestrial life and is chosen to make 'first contact'. The story was written by Carl Sagan and Ann Druyan. While experiencing the interminable wait in the on/off Hollywood production schedule, Sagan adapted the film into a novel.

As creative director of the Golden Record, Ann was responsible for one of our more tangible attempts at communication with

---

* I call it 'teenage loneliness': that may be where it starts, but it may not be the age at which it finishes.

aliens. The two records contain images and music that were chosen to represent life on Earth, should an ET intelligence discover either of the *Voyager* spacecraft. The Golden Record is not merely a good introduction to Earth for extraterrestrials; it's a good introduction to Earth for terrestrials. On the first tour of science that I went on with Brian Cox in 2011 we played the music from the Golden Record in the auditorium as the audience found their seats. There were works by Chuck Berry, Stravinsky, Bach, the Navajo Night Chant and the Mahi musicians of Benin, as well as much more. As the audience listened, it was a delight to peek through the stage curtains and see their faces changing from confusion to intrigue, and then the realization that they were listening to what might be an alien's introduction to our planet. There was a full transformation from puzzlement to delight.

The Golden Record is like a playlist made for a hoped-for lover. It says, 'Here is some music I like, I hope you like it too. If you do, then maybe we have something in common, and maybe we can go for a date somewhere near the Pleiades.' It is also a reminder of our parochialism. For many of us, nearly all the music and art we experience lies within the boundary of what is considered our culture. There is a vast amount of creativity that is considered specialist or arcane because it is beyond our national boundary – a border that can encircle imaginations.

NASA was not initially keen on the idea of a Golden Record, and others complained about the playlist. For a journey that was to be so expansive, some of the imagination behind it was at times very parochial. Some complained that the Golden Record needed more Frank Sinatra, instead of what they saw as marginal world music – 'Alima Song' by Mbuti of the Ituri Rainforest, and 'Naranaratana Kookoo' ('Cry of the Megapode Bird') by Maniasinimae and Taumaetarau Chieftain Tribe of Oloha and Palasu'u Village Community. Blind Willie Johnson, Louis Armstrong and Chuck Berry weren't enough to sate NASA's nationalistic desires. Ann remembers how unsettled NASA executives were the first time the record was played, when they started hearing the Japanese gamelan music and all the other different musical traditions. Carl Sagan stuck by the decisions of the Golden Record team.

Musical taste often seems barely to travel across a national border, so it's hard to say how it will fare beyond the boundary of our solar system. It seems likely, though, that nothing will ever intercept either *Voyager* craft, as space is just too big and too empty, and *Voyager* is small enough probably to fit in the room you are sitting in now, but it is fun to imagine the possibilities.

For Ann, connecting with alien life forms is not simply a problem of vast distances, but also of time. We need our alien intelligence to be broadcasting, and we need to know how to receive. For 4.5 billion years we could have been bombarded with radio signals and we wouldn't know a thing about it, because those signals would be totally invisible to us. The distances these signals are likely to be travelling means that the time taken to receive them may be so long that a civilization has fallen by the time we hear its first words.

Ann's view is that if you have as many stars as we have in our galaxy, and as many worlds, intellectual development will not end with the technology that we have now, with radio. There are other possible methods of communication that we haven't perceived or stumbled upon yet. She spends time each day looking at her Instafeed, exploring the different forms of life on Earth in wonder, and as she contemplates what evolution has produced in our environment, she ponders what shapes and forms life might take on other worlds. Will they be ugly to us? Will we be ugly to them? Will we recognize each other's intelligence? Or will our new alien friends merely want us to tell them where they can find Chuck Berry?

Have you got the patience?

Of all the attributes needed in the search for extraterrestrials, patience is probably the most important. Astrophysicist Frank Drake, who performed the first SETI experiment in 1960, recalled that Carl Sagan eagerly walked in to listen for signals, but after thirty minutes, having failed to detect transmissions from other life, he had fallen asleep. Like the brief certainty you feel that your lottery numbers are going to come up today, the likelihood that today probably won't be the day kicks in very quickly.

Seth Shostak has been waiting for a sign for a long time now. He is the senior astronomer at SETI, which scours the sky seeking signals that would suggest the existence of a distant creative imagination that is the product of biological life, rather than an object obeying only the laws of physics. Like many of us, Seth's introduction to the idea of alien intelligence came from the B-movie science-fiction films of his youth, such as *Earth vs the Flying Saucers*, *When Worlds Collide* and *It Came from Outer Space* (which I have been privileged to see in its intended 3D format, ducking as the opening-credits comet hurtled towards me). Seth was also a regular visitor to the New York planetarium and built his own telescope when he was ten years old; but it was when he was a grad student using radio telescopes in California that his destiny was made manifest.

He was based at the observatory behind the Sierra Nevada mountains. During the day there were other people at the observatory, but at night he was the only one there running the telescope. At the observatory library he picked out *Intelligent Life in the Universe* by I. S. Shklovskii and Carl Sagan. When Seth read that book, he realized that the instruments that were grinding away outside the window where he was sitting could also pick up signals from extraterrestrials, and the excitement of the possibilities immediately struck him.

Having now spent so much of his life trying to detect alien life, what does Seth think the public reaction would be, if strong evidence was found? He says there has been very little research into this question, despite it fascinating many people. If he gives a talk about SETI, that is the question that always comes up. In the USA, home to the secret Illuminati headquarters where all the conspiracy theories are made, Seth tells me that many people believe that such a finding would be covered up (if it hasn't been already, obviously) and that, whatever was told, it would be seen as a lie covering up something else. He reckons a cover-up of such a discovery would be impossible. Even the news of false alarms gets out fast, with *The New York Times* calling him within eight hours of one particular such episode. As far as Seth is concerned, though, 'The public wouldn't go nuts. They wouldn't riot in the streets.'

In fact when they picked up a signal that was later refuted, nobody seemed to care. In 1996 there was a story about the possible existence of Martian microbes, which people found interesting and which made front-page news, but everybody went to work nevertheless. Martian microbial life did not disturb terrestrial commuting.

Equally, if we do pick up a signal, it will be coming from very far away. You can't start chattering away in return. The aliens might be dead by the time you reply to their question 'Have you found the secret of perpetual motion?' Or 'We just wondered if we left our pyramids there last time we visited?' Seth believes that the consequences of connection would take effect in the long term, in the same way that the discovery of America by Europeans had very little immediate consequence, but ultimately the opening up of the western hemisphere had major consequences.

Will the discovery of other life change our sense of meaning? Not for Seth, as he already believes there must be intelligence out there; but it would still change his personal philosophy somewhat, if extraterrestrial life forms were discovered in his lifetime. He compares his job to being a cancer researcher. You can be sure there's a way to beat cancer, but whilst it doesn't mean you will be the one to find that way, your work might be one of rungs on the ladder that others will use to get to that eventual goal. He may spend his whole life seeking extraterrestrial communication and never hear or witness anything, but it doesn't mean the adventure was worthless.

Seth finds optimism in our ingenuity, and hope in our machinery. 'The real point is where is humanity going? It could be that we're going in the direction of self-driving cars, extending lifetimes to 300 years or whatever. And we're moving in the direction of inventing our successors, which is to say, the thinking machines.'

If humanity is about to come up with something better – machine intelligence – then machine intelligence will not be stuck on a particular planet. It may not be stuck with a one-million-years species lifetime, either. It reminds Seth of the science-fiction classic *Forbidden Planet*, a spaceship reboot of Shakespeare's *The Tempest*. On the planet of the Krell, the crowds are gone, but the machinery is still there. Seth thinks that changes the picture fundamentally, because we continue to think of aliens as biological and, if they are not biological, then applying the kinds of scenarios that biology is tied to may be wrong. In *Star Trek: The Motion Picture*, perhaps the most philosophical of the film series, the intelligent life that they communicate with, V'ger, turns out to be a *Voyager* space probe that has become sentient.

Seth knows that we are 'social critters', so why stop your socializing within your own atmosphere? He likes to assume that there is a wide spectrum of aliens out there. Some might be less advanced than we are – the equivalent of jungle apes – and a chance to meet a form of our evolutionary ancestors. Or they may be very much more advanced. He thinks that the chance that they will be within 100 years of our own level of development is 'very small'; and maybe even within 100,000 years is unlikely. They might be thousands, millions or even billions of years more advanced. It is not going to be the cantina scene in *Star Wars*.

Seth is not expecting much difference between Earthly and alien microbes, either. A microbe's form is determined as much by physics as by the specifics of its biology, needing only the minimum surface area for its volume and chemical needs. Seth reckons that microbes are going to look superficially the same, that you are going to need a microscope to see them and that they are going to be little round or elongated blobs. You're going to have liquids, because otherwise the chemistry doesn't go anywhere. It is all down to how long biological intelligence can exist. Will it exist long enough to create the machine intelligence that will increase the chances of longevity? Machines can beat Darwinian evolution. You can unnaturally select for a rapid increase of memory, or whatever offers you the advantage for a less-finite existence and a rapid sharing of knowledge accrued. To Seth, biology might just be a coming attraction.

In the gloomier analyses of the Fermi paradox, civilizations advance to the point of their self-willed destruction. In Seth's vision, perhaps we won't destroy ourselves, but instead we will remodel ourselves and supersede nature. Seth keeps himself abreast of the latest portents of doom for humanity, but when he analyses the things that might destroy us – including pandemics – the only thing that has any significant impact would be if you let all the nuclear weapons fly. If the biggest bomb went to the biggest city, and the second-biggest bomb went to the second-biggest city, then eventually you would run out of bombs at somewhere around the size of Oxford; and, assuming a 100 per cent kill rate, you would get rid of one-third of all humans. I hope that is right, though I am not sure it takes into account death by collapse of the infrastructure, or radiation sickness, or the new tribes of aggressive cannibals that usually turn up in similar apocalyptic-movie situations (though, as I mentioned previously, the optimistic view of being a meal is that you have found purpose, and your purpose is to be delicious). Seth believes this is the worst outcome for humanity, short of an asteroid impact or some natural disaster. If you're leaving it up to humans, they can do a lot of damage to one another, but they can't get rid of one another. It is a strange note of optimism when considering all-out nuclear war.

Does Venusian trip off your tongue?

The amateur astronomer Patrick Moore, who was England's greatest ambassador for stargazing, as well as being a keen xylophone player, made a documentary for the BBC series *One Pair of Eyes* about people whom he described as being of independent thought – people who were not shackled by the chains of convention.

His subjects were observers of UFOs and flat-Earth enthusiasts. Moore generously celebrated their eccentricity, little knowing that fifty years on, these niche passions would have generated a dangerous mainstream of crackpottery. Mr Moore meets Mr Roberts from the Aetherius Society, founded by George King, a man in communication with a Venusian by the name of Aetherius – a Greek pseudonym, we are told, meaning 'a man who comes from outer space'. Mr Roberts believes that humankind used to inhabit a planet between Mars and Jupiter, but in humans' deviation into material science, they blew the planet up, and that is now observed as the asteroid belt. There is also some room for marauding fish people. Saturnians are a large ovoid shape, perhaps forty feet in diameter. 'I can speak altogether three of the space languages – one is Venusian, the second is Kruger... and Pluto,' says Mr Roberts.

To speak these languages, Mr Bernard Byron believed that he had them sent to him by rays from the planets. The 'language' is a delightful spectacle, like different forms of gobbledegook somewhere between a tic-tac man on the racetrack with hiccups and the vocal track of a Norwegian Eurovision Song Contest entry played in reverse. He also writes down some Plutonian and Venusian language. Will the newly discovered phosphine of Venus prove that Mr Byron's linguistic gymnastics were true after all?

As Carl Sagan commented, for all our hopes of being able to cross the language barrier with aliens, we must remember that we have not yet found a way to effectively communicate with the intelligent creatures of Earth, such as dolphins and whales. For the time being, the intergalactic language of choice is mathematics, although physicist Brian Greene occasionally worries that the aliens might arrive one day and ruefully observe, 'Ah yes, we used to think the language of the universe was mathematics once, too.'

Brian considers that one of the greatest discoveries of the last ten to fifteen years is that most stars do seem to have planets in orbit. In our galaxy alone, it has been estimated there could be tens of billions of planets, and a significant fraction of those might have conditions that would be hospitable to some form of life that we can imagine existing. Brian knows that's not enough to give any predictions about how likely it is that there's life in the universe, though he adds, 'I do think it's likely, just because I have the intuition that I think many people share. There's so many opportunities for life to form out there. And, therefore, there's a sense that it must have happened to somewhere else as well.' Life is one thing, but intelligent life is quite another. Perhaps, as we might be discovering, intelligence is a brief evolutionary advantage that can go awry.

Brian would consider discovering a planet that 'just' has bacteria to be a monumental discovery. We could compare it to life here on this planet and it would help us shape our conjecture a little. After all, much of what we are made of is microbial life, so it would be a good start.

Intelligent life is the really deep and interesting question for scientists like Brian. We have no idea of the likelihood of intelligence forming. Maybe it was simply a lucky accident that an asteroid slammed into our planet and wiped out the dinosaurs, and that enabled mammalian life to develop and dominate. Brian wonders if, without that asteroid, there would still be dinosaurs walking around? 'Maybe at this point they'd be talking to each other and conversing, having set up countries and businesses and commerce.'

'I'm a doctor, not a bricklayer' – Dr 'Bones' McCoy[2]

For the time being, we haven't been able to signal our existence to others any great distance from us. We've been waving by radio signal, so we require the extraterrestrials hopefully to be fans of *The Jack Benny Program* or *The Archers*, if we want a response. There is a little shell around Earth of eighty light years, in which the totality of every signal that we have sent into space is currently constrained. If the extraterrestrial intelligence is looking for intelligent life elsewhere in the cosmos and they are trying to find out whether Earth has it, they will have to be within that shell of eighty light years; and if they are further away, then those signals will not yet have had time to reach them. If they do encounter that signal, well, it's going to be maybe eighty light years before we would get their response.

Tim O'Brien had the keys to the gate of the Lovell Telescope on the day I visited Jodrell Bank, so I am indebted to him for the chance to take the lift and the ladder into the centre of the dish. He has also played a major part in the Bluedot Festival. Tim has combined these two disciplines of music and astronomy by fusing music and sounds detected by the telescope, such as pulsars and the bleeps of space missions. Collaborators have included bands such as Sigur Rós and New Order. Tim created a single, 'Hello Moon, Can You Hear Me?' with Tim Burgess of The Charlatans. 'Hello Moon, can you hear me?' was a sentence that was bounced off the Moon in the late 1960s. Bernard Lovell himself was not originally keen on using his telescope to search for signals of extraterrestrial life, considering it a mere frippery for such technology. He changed his mind.

As with almost everyone else, Tim's initial interest in aliens stems from science fiction. For him, it was *Doctor Who* and *Space: 1999*, in which aliens with eyebrows made of hundreds and thousands could transform themselves into falcons and tigers.* His fascination has become more factual over time, but it was an area of interest that could be easily dismissed in the past. At a conference in 1985 no one was studying exoplanets – planets outside the solar system – and it would have seemed a very odd thing to do. This has all changed now, and it is a very important part of modern astronomy.

Like Seth Shostak, Tim believes there must be life elsewhere, possibly even in our solar system, such as microbial life on Mars. It seems too weird for life to have evolved only here. Optimistically,

* To be honest, I cannot remember if they could turn themselves into tigers. I might be thinking of the short-lived TV series *Manimal*.

196

there should be at least a few other civilizations in our galaxy; 200 billion stars offer a lot of possibilities for quite a few of them to hold an Earth-like planet or two in orbit.

Tim believes that if every other planetary system had some sort of technological civilization on it, then we would have seen evidence by now. He also has no fear that communication with aliens will lead to our destruction, believing that if they were bloodthirsty brigands, they would have found us and slaughtered us already. 'That's my optimistic view,' he says, 'that there are nice civilizations out there that might be helpful to us.' Unsurprisingly, the search for extraterrestrial life seems to be the territory of the optimist. Tim describes us as a little weak species, but one that is on an upward trajectory. There are bumps and dips and troughs in the road, but he still thinks we are learning, and we probably know what the right direction is.

Ever the optimist, Tim believes that by the time we do get to the point where we have developed technology that will allow us to travel large distances, even between stars, we will also have developed ethically and intellectually enough to deal with extraterrestrial communion in whatever form – and it may be one very different from our own – it takes. Tim is reminded of Q in *Star Trek: The Next Generation*,* an extra-dimensional being with power over the laws of physics. In the original series of *Star Trek* a mining planet is found to be home to sentient rock. If the imagination of nature is greater than the imagination of human beings, then who knows what shape or form (or formlessness) some alien intelligence that is thoughtful and alive will consist of?

* Yes, yes; and *Deep Space Nine* and *Voyager*.

Tim imagines a possibility of a being that would understand the physical world, the biological world, and would understand that there might be things alive that they have to avoid harming. During the age of exploration, explorers did their fair share of deliberate slaughter, but also brought the accidental slaughter of diseases for which indigenous people's immune systems had no defence. Perhaps some intergalactic species will stay home, for that reason.

If it did turn out we were alone in the universe – and we have a lot of ground to cover before we could even have an inkling of how likely that really is – Tim would be disappointed. It would be a pity not to have the chance of meeting a truly alien species with a completely different perspective, and meeting all those challenges of communication. If we are alone, it should be a wake-up call, though. The rarer life seems to be, the more precious we – and the complex living things around us – become. It is quite a responsibility to be rare, especially when you have at least some control over your possible destiny.*

Does Tim ever have a moment when he thinks: This time it really is LITTLE GREEN MEN!... OR LITTLE GREEN WOMEN!... OR AN EXTRA-DIMENSIONAL BEING OF MANY GENDERS... OR A THOUGHTFUL PIECE OF LIMESTONE?

There are shapes and patterns to expect, when looking at radio signals from events such as the explosion of stars, but if you decide to look at this in very fine detail and zoom in on it, and look at the little peaks and troughs, you can see all manner of weird signals

* With the usual codicil of free will being an illusion.

that appear and disappear, and very narrow band-things that are basically going to be signals from technology. At this point you could find yourself exclaiming, 'Oh my God, what is that?' Tim is trained not to leap to the exciting conclusion that it's an alien sending a message. The necessary scepticism leads to the question: what's the most likely thing to be producing that? Inevitably it's always most likely to be the microwave oven, or the laptop, or the mobile-phone signal. Those signs of alien life really are terrestrial, but at least there is the solace that they lead to a reheated vegetable biryani in the kitchen.

There was one night when cold, hard scientific reason was usurped by the adrenaline of expectant delight. Tim was working on an observation with an optical telescope, looking at the remnants of exploded stars from a telescope in the Canary Islands. It was just him and a PhD student on the top of a mountain, amusing themselves making observations of exploded stars around the galaxy. They were looking at things they knew had exploded in the past. People had seen this bright new star up in the sky and then it had faded away. They were taking deep, long-exposure images of the area of sky where it had been shown to have appeared in the past. As they were looking for evidence of the expanding remnants of the explosion, they found an amazing, bright sort of nebulous object alongside one of the stars. Measuring the speed at which it must be expanding, they calculated that it was at huge speed. They were quite excited, calculating things like how much energy must have been involved in the explosion. As they started to do more observations of some other bright stars, they noticed one next to another one. And then they noticed one next to another star. The excitement mounted.

Then they went outside, only to realize that there was a light leak at the side of the camera on the back of the telescope, which was reflecting light into the optics, and that what they were seeing was merely a reflection of the star itself in the image. The great new discovery was extinguished by use of some gaffer tape and a few big black bin bags, but there had been a good hour or two of excitement at their amazing discovery and what it might mean.

Astrophysicist Frank Drake discovered what has come to be known as the Drake equation. It collates all the astronomical considerations for there to be the development of complex, communicating life in other parts of our galaxy, such as the rate of star formation in our galaxy, the number of stars that have planets and the length of time it might take a civilization to start emitting detectable signals of existence. Drake has spent much of his life listening out for signals from beyond the Earth in the hope of detecting life, but he has also devoted some of his time to listening out for the loneliness of those of us who are trapped here.

Seth Shostak worked with Frank Drake. He told me that he would walk down the hall and into Drake's office and, however busy he was, Drake would always stop. He would turn and say, 'Hi, Seth', and he was always ready to listen to him. Seth pointed out to Drake that he didn't really know anybody at the institute who was that willing to interrupt what they were doing at any time, to listen to somebody who just walked into their office. He asked him, 'Frank, tell me: how do you do that? Is it because you have children?' And Drake replied, 'No, it is because I have students.'

In *A Slender Thread*, Diane Ackerman wrote of how Frank Drake demonstrated his interest in life, both terrestrial and extraterrestrial. She wrote, 'With his prematurely white hair and gentle manner, he looked like someone distressed souls could turn to, and they did.' He took training and spent nine years working on a crisis helpline, talking to people who were contemplating suicide and feeling lost in the world. There is something beautiful in that, although he was professionally focused on the enormity of the universe, Drake did not lose sight of the life that he knew existed here on Earth. When we ponder ourselves being alone in the universe, we must never lose sight of the fact that there may be life that we know exists, which may need us to reach out and communicate with them, and they could be just down our street. While we wonder if we are alone in the universe, there may be someone who is looking out of their window very near you and thinking, 'I am alone.' Time to make contact.

> You're an interesting species. An interesting mix. You're capable of such beautiful dreams, and such horrible night-mares. You feel so lost, so cut off, so alone, only you're not. See, in all our searching, the only thing we've found that makes the emptiness bearable, is each other.

A MESSAGE FROM AN ALIEN IN *CONTACT* (1997)

CHAPTER 7

Swinging from the Family Tree – Inviting Yeast to the Family Reunion

Take your stinking paws off me, you damn filthy ape.

George Taylor in *Planet of the Apes* (1968)

The original film of *Planet of the Apes* was scripted by Rod Serling, creator and narrator of *The Twilight Zone*. I adore his work. His imaginative fiction was always brimful of humanity, kindness and empathy, and frequently that meant it had a political edge, too. *Planet of the Apes* can be seen as a film about slavery and how we project inferiority on others, condemning them to a lowlier status. When it was released, the USA was witnessing increasing civil unrest and the inspirational activism of Martin Luther King, with whom Charlton Heston stood on the steps of the Lincoln Memorial, declaring him a 'twentieth-century Moses'. King was assassinated the year that *Planet of the Apes* was released.

Now, the film may be seen in an even broader context – not just as an allegory about division within a species, but also more directly about our judgements and presumptions concerning other species, their sentience and their ability to feel pain. We are in a tribe that grasps for exceptionalism, and we don't take well to it being diminished by the achievement of other species.

When Darwin published *On the Origin of Species*, numerous cartoonists mocked the very idea that we might be connected to those dirty apes. More than 160 years on, there are still multitudes of people across different faiths and nations who mock the idea of our connection with 'brute nature' and, ironically, other opinions that they share are frequently inhuman and brutal. But our connections are now inescapable. We are apes. And we are chemistry. And we are information. Much of that information and chemistry connects us to all the other species.

Inside us is a museum of life – strands that tell the story of how life came to be, and how it took shape and form. Structures as strange as life don't appear to be occurring with great regularity across the universe, and yet I've spent too much of my life blasé about its rarity. With that rarity comes absurdity. To be alive is to be exceptional. Our wish for extra exceptionalism has encouraged us to disconnect from the rest of nature. We float above it – our sense of superiority ignoring all the living things, microbes and bacteria that are dancing about, in and around us, as we attempt to be aloof. The idea that our connections to life are an insult to our humanity should be insulting to our intelligence. To me, the insult to humanity is to be blind to a reality because it hurts our egos. How silly to be so frail.

Through methodical field studies and breakthroughs in genome sequencing, it takes aggressive ignorance to maintain that our position as human beings is unique and somehow unattached to the Tree of Life. Not only does our genetic code show indisputably close links, but observing our social behaviour – and the social behaviour of the species nearest to us – shows up many uncanny similarities. I think it is enriching to realize that chimpanzees and human beings have travelled the same ancestral path for much of our history.

The sense of being God's chosen species is sullied by the sight of a chimpanzee's pink bottom, or a bonobo ape's sexual shenanigans. Creationist politicians in the American south have garnered hearty laughs and applause by mocking the notion of a shared ancestry with the other apes by doing their best 'Look at me, I am moving like a chimpanzee' walk. Unfortunately, science shows that a human walk is not markedly different. I think understanding our deep connections with all other species makes every story of life more valuable. But then how do we convince those who reject our life story, and should we even bother?

The Creationist movement believes that our lives were made more special than the rest by God. Frequently that belief accompanies other literal beliefs in Bible stories, allowing us the right to abuse other animals (and sometimes people), however we wish. Creationist funding has led to a rebooting of Creationism as the theory of intelligent design. Key proponents have spent a great deal of time and money obsessing over the flagellum of bacteria – the appendage that can assist their swimming –

seeing it as one of the key pieces of machinery that must have been designed by a creator, rather than something developed through mutation, heredity and natural selection. The argument has been that the flagellum shows irreducible complexity; that there were no intermediate stages – it was just there, manufactured by a maker.

There have been many rebuttals of this presumption by scientists. Cell biologist Kenneth R. Miller has often written about the flagellum, and has frequently criticized and debated intelligent design and Creationist proponents. He is also a Roman Catholic.[1] Miller has written, 'I do not believe, even for an instant, that Darwin's vision has weakened or diminished the sense of wonder and awe that one should feel in confronting the magnificence and diversity of the living world.'[2]

The flagellum is up there with the banana[*] as one of the key exhibits that refutes evolution by natural selection. If you get talking to an intelligent-design proponent, you will hear a lot about bananas and bacteria: that these were the works their God really sweated over, and that the rest is mere footnotes. The view now seems to have become truly embedded in mainstream US right-wing politics, with the startling anti-evidence extremism of the Republican Party, a party that would now make even Richard Nixon blush. How do you prise people's hands away from ludicrous theories of life on Earth that lack the richness and connection of the real story of life?

[*] Bananas are used because they are seen as the perfect design for us to eat, with their 'ring-pull', easy-to-handle shape and stout packaging. That the banana we know now is the product of artificial selection must not be mentioned.

In 1981 there was a challenge in Arkansas to the state's crea-tion-science law that demanded a false balance between the teach-ing of evolutionary science and creation science – which is less of a science than big fiery-dragon science and werewolf science, both of which are considerably more interesting, too. Carl Sagan travelled to Arkansas to testify as a friend of the court. He spent an afternoon calmly explaining the differences between science and creation science, never losing his cool and never sneering at the questions or presumptions. He didn't mention dragons or werewolves once.

A year later Sagan received a lengthy handwritten letter from the creation-science expert who had been testifying for the other side. He was now a former creation-science expert. He thanked Sagan, writing that Carl was 'humble and kind' when he asked his questions, and that really disturbed him. He concluded, 'I had to do this self-examination and I want you to know that I'm getting an education degree now, so I can teach biology. If you hadn't been so, so humble, I don't think that doorway would have opened for me.'

It is something Ann Druyan often thinks about. 'Too often the people who try to present this view of the world are feeling a certain kind of satisfaction in being smarter than the people they're trying to change.' She thinks you have to show profound respect for everyone, and be able to look them in the eye deeply and listen to what they are saying and understand their motivation. In all the time that she knew Carl, she says she never heard him speak to impress anyone with how much he knew – only to communicate.

Creationism fears the loss of exceptionalism. Less religious people – even godless scientists – have worried also about the loss of human exceptionalism. As researchers have gone deeper into the wild and spent time observing other species, some of their discoveries of other species' intelligence and sense of society have been coldly received at first.

'I wanted to come as close to talking to animals as I could – to be like Dr Dolittle. I wanted to move among them without fear, like Tarzan' – Jane Goodall

Naturalist Eugène Marais wrote two fascinating books on the work he had done closely observing a single species, *The Soul of the White Ant*, which – oddly – is about termites, and *The Soul of the Ape*, which – oddly – is about baboons. Other than the misleading titles, these are fulsome and intricate examinations and are groundbreaking in their approach. In a letter to a friend, he wrote, 'No man can ever attain to anywhere near a true conception of the subconscious in man who does not know the primates under natural conditions.'[3] He also wrote a paper on baboons that was published in his lifetime, called 'Baboons, Hypnosis and Insanity',[4] but he had been dead for thirty-three years by the time *The Soul of the Ape* was published in 1969, by which time Jane Goodall was nearly a decade into her pioneering, deeply focused studies and research on chimpanzees.

It was a cartoon that first drew me to the work of Jane Goodall. The educational possibilities of cartoons are, in my view, often overlooked, as anyone who has learnt about existential anxiety via Charles Schulz's *Peanuts* strips can testify. In one of Gary Larson's *Far Side* cartoons, two chimpanzees are seated on a branch. The female is angrily confronting her partner as she grooms him. 'Well, well, another blonde hair. Conducting more "research" with that Jane Goodall tramp.' The Jane Goodall Institute took offence and Larson was worried, as he had deep respect for Jane Goodall and her contributions to primatology. Fortunately, Goodall herself found it funny and a friendship blossomed.

The awe that surrounds her is deserved. Before Jane, chimpanzee research had focused on understanding primates in captivity. It would be as if the only valid study of humans was to observe them in maximum-security prisons. She received much criticism when she started a dialogue with scientists who did imprison apes. She wanted to get the conversation about vivisection out in the open. She believed that protesting was not enough to bring change and, like Sagan with the Creationists, she knew that she had to engage, and persuade those who dismissed her to come round to her point of view. This led to her making visits to laboratories where chimpanzees were being experimented upon, which were deeply disturbing. On one visit she was filmed looking at an elderly chimpanzee in a spartan cage. They seemed to gaze into each other's eyes silently for some time. Eventually Jane began to cry. As a tear rolled down her cheek, the chimpanzee gently reached out and wiped the tear with her finger. Jane has talked of rare moments where she has felt there has been 'mind-to-mind' connection between her and a chimpanzee, and watching this particular moment, it certainly feels profound.

That connection between human and animal reminded me of truck driver Rick Swope, who risked his life at Detroit Zoo in 1990 when he saw a chimpanzee called Jo-Jo drowning. When asked why he took such a risk, he said, 'Well, you see, I happened to look into his eyes, and it was like looking into the eyes of a man, and the message was, "Won't anybody help me?"' To the press at the time he said, 'It was no big deal, you know. It wasn't nothin' that hard. It didn't take an exceptional person to do it. If it did, I couldn't have done it.'[5]

David Greybeard was the first chimpanzee Jane Goodall built a relationship of trust with. Again, it was in the eyes. Jane was following David and she picked up a ripe palm nut and held it out in the palm of her hand for him. He turned his head away. David didn't want it, but Jane felt a bit cheeky. 'I pushed my hand a bit closer and he looked directly into my eyes and, whilst looking into my eyes, he took the nut and then dropped it and then very gently squeezed my finger. This is how chimpanzees reassure each other, so we had that perfect communion without words, a method of communication that pre-dated words.'[6]

Dr Goodall's research included the discovery of how chimpanzees had developed the use of an effective tool for catching termites. When the famous and revered paleoanthropologist Louis Leakey received an excited telegram from Jane about observations of tool use, he famously responded, 'Now we must redefine tool, redefine Man, or accept chimpanzees as humans.'

Jane's research did not follow the academic rules of engagement. She believed that chimpanzees possessed personality, and she dared to give them names rather than merely number them as specimens. Her system was a maverick one of scientific engagement, because she was not academically trained in the rules of the science of the time. This lack of cold, objective distance, this ability to imagine the minds of the chimpanzees, to empathize with them, to break down that barrier, to feel for and with the chimpanzees of Gombe, brought with it incredible understanding.

The emotional cost can be seen when we watch the devastation of the Gombe chimpanzee community during a polio epidemic. Jane considers it to be the worst time she has ever lived through. By daring to imagine the fullness of their internal lives, the pain was all the greater when she saw the chimpanzees that she had become attached to struggling with atrophied limbs, many of them dying.

There was also the revelation of brutality within the community. Once, Jane had thought that chimpanzees were 'like us, but nicer'. Then she observed Passion and Pom, a mother and daughter, who slaughtered and cannibalized at least two babies. There was also a brutal and devastating tribal war. For those fearing the loss of the positive sides of human exceptionalism from Goodall's close observations, there is perhaps the cold comfort of knowing that pointless violence and murder are not merely some grotesque aberration reserved for human civilization.

To see and feel the link between the human and the chimpanzee – not merely the link of genes or branches on the evolutionary tree, but the link in our chains of thought and behaviour – to see the connectedness, illuminates the whole picture of life on Earth. It dares to add emotional content to the rigour and calculations of what we think of as typical science. As Goodall says in the documentary *Jane*, 'Staring into the eyes of a chimpanzee, I saw a thinking, reasoning personality looking back at me. I felt very much that I was learning about fellow beings, capable of joy, sorrow, and jealousy.'[7]

Jane believes that her lack of a university education enabled her to approach her research without being preconditioned into using 'accepted' methods. When she went to Cambridge University, she experienced hostility, as her work was considered not 'detached enough' to be science. Jane saw this as utterly wrong, because 'If you have empathy you see something a little strange and you just feel what's motivating the chimp and that gives you a standpoint from which you can then examine "am I right or wrong?"' She was amazed that other scientists felt you couldn't talk about individual chimpanzees and that you should only focus on one aspect of their lives, such as their feeding habits. At this point, nobody had really studied anything at all about wild chimps in Africa, and Jane couldn't believe she was being told that she was supposed to pick on feeding or grooming or 'something stupid'.

Were the results of her research treated dismissively because they interfered with our story of exceptionalism? She was told firmly that the difference between us and other animals was 'the difference of kind, not degree' and that animals did not have personalities, minds or emotions.

One of Jane's childhood inspirations was Hugh Lofting's *Dr Dolittle* stories, all about the doctor who could talk to the animals. The doctor is taught to talk to the animals by his Polynesian parrot. Unfortunately, in interspecies linguistics such shortcuts have not been found, although in the latter part of the twentieth century some researchers tried to teach other species our way of talking. With chimpanzees like Nim Chimpsky, it was via the use of sign language, while John C. Lilly and Margaret Howe Lovatt hoped to teach a dolphin[8] called Peter to verbalize through his blowhole. Neither story ends happily, either for the research results or for the animals involved.

Bleakly, the story of Peter the dolphin potentially illustrates the darker side of animal intelligence, as some people believe Peter took his own life. Having been in a reasonable-sized pool and with a close relationship to Howe Lovatt, he was then moved to a dark and very small pool and left alone. One day Peter took a breath and then sank to the bottom of the tank and made no attempt to surface again. For some, this was believed to be a dolphin suicide. Ric O'Barry, an animal-rights activist who previously caught and trained dolphins, believes he saw Kathy, one of the dolphins from the TV series *Flipper*, commit suicide. 'The suicide was what turned me around. The [animal entertainment] industry doesn't want people to think dolphins are capable of suicide, but these are self-aware creatures with a brain larger than a human brain. If life becomes so unbearable, they just don't take the next breath. It's suicide.'[9]

We have realized that many other species require space and stimulation. Emotions and needs that we have considered unique to ourselves have turned out to be requirements of many more species.

Jane Goodall has observed much non-verbal communication in chimpanzees that is comparable to the way humans communicate. Chimpanzees communicate with postures and gestures that include kissing, embracing, threatening, swaggering, fist-shaking and rock-throwing. She says she has noticed the human-like swaggering and posturing especially in male chimpanzees: 'It's just like some male politicians that shall be nameless.'

More recent research in the field, by primatologists such as Cat Hobaiter, has seen that the physical communication between chimpanzees is even more complex than previously imagined. A chimpanzee wishing to tell another to 'get down from that tree' might shake a branch while scratching with the other arm. If ignored, that gesture might become more dramatic, with a more vigorous shaking of the branch and the other arm raised; this is 'You bloody get down from that tree right now or there's no tea for you.' Cat's research tells us that there is even more going on, as the full meaning is 'get down from that tree' (the shaking of a branch) and 'come and groom me' (the movement of the other arm). This does not suggest such a complexity of communication that *Macbeth* is on the cards, but perhaps there is enough to run a moderately-sized beauty salon. Rather than belittling our own achievements in language, these breakthroughs in understanding seem to offer the increased possibility of interspecies communication. So many things that may seem to reduce our significance ignite our possibilities. I think the ability to discover such things is more significant than the belief in a single 'superpower of spoken language'.

One of my favourite stories from Jane Goodall's observations was how she witnessed the excited celebration and curiosity when a group of chimpanzees discovered a waterfall. She had already seen what she considered to be rain-dances, but to her this waterfall encounter was 'similar, but different'. It was wilder, something she describes as 'an answer to the elements... as they approach the waterfall and hear the thunder of the water hitting the rockiness of the stream bed and the hair starts to bristle, they sway back and forth and sometimes hoot'. Unusually, the chimpanzees actually entered the stream and threw rocks, when normally they never get their feet wet. For Jane, the most amazing part came after these first celebrations. The chimpanzees sat on a rock and their eyes followed the water as it fell. She wondered about how, if they had words, they might share with each other what they felt, and whether or how that would have further transformed the experience? She imagined them thinking, 'Why do I feel this way? What is this magical stuff that is always coming, always going and always here?' She sensed an element of ritual, as if she was observing what might have led to the early animistic religions – the worshipping of things not yet understood, like the Sun and the Moon.

Jane sees the words that triggered this explosive development of our intellect as the biggest difference between chimpanzees and humans, but in other ways she realizes that we are so similar. The chimpanzees are far more intelligent than people used to think, but Jane adds, 'We have designed a rocket that has been up to Mars and a robot crawled about it, and think how you and I are talking now. That's the big difference, and how stupid it is that the most intellectual creature to ever walk the planet is busily destroying its only home.'

We end our conversation talking about a pig called Pigcasso, which has gained celebrity due to his painting skills. Jane has found that anyone who watches Pigcasso in action can never eat bacon again. Brian Cox had a similar experience. He is usually an energetic omnivore, but he no longer eats octopus. During a diving adventure, he ended up communicating with an octopus. He left the exchange persuaded of their rich intelligence. If you are an animal-rights activist in need of a convert to vegetarianism, all you need to do is start bringing your most intelligent painting pigs, dancing cows and haiku-reciting hens to Brian's house and, after a quick session, you will have a new vegetarian on your hands.

The careful observation of animals in the wild has broadened our perspective and increased the potential of interspecies empathy. Jane has written about how wary we must be of the limitations of our perspective; that we peer through only one window, so 'no wonder we are confused by the tiny fraction of the whole we can see'; and that we must learn to find a way to reach other windows. As she tells audiences around the world, 'Only if we understand, can we care. Only if we care, we will help. Only if we help, we shall be saved.'

But what about empathy with hedgehogs?

'I will be down on my knees sniffing poo'

Jane Goodall was Hugh Warwick's third childhood crush. The first was Maid Marian in the form of a cartoon fox, from Disney's animated *Robin Hood*, and the second was Kate Bush – so all very suitable for someone who grew up to be a naturalist.

Now Hugh has a vivacious enthusiasm for the natural world – and not just the things we might obviously find attractive. For Hugh, scientific understanding not only doesn't remove the beauty of the flower, but it can make beautiful things we might consider disgusting, and he'll shove your face in it to prove his point. When Hugh sees otter excrement, he drops to his knees and inhales deeply and insists that his walking companions do the same. According to Hugh, it smells fantastic: of jasmine and Earl Grey tea, with maybe a very slight tang of digested fish, but you can get over that.

Like Jane, Hugh has had moments of connection that go beyond mere observation. It first happened when he was living in

216

a caravan in a field and radio-tracking hedgehogs. Once a week he would bathe in a local farmhouse and make the loaf of bread that would sustain him. He rarely smelt of jasmine and Earl Grey, except maybe on a Monday. In isolation, he found that he was talking to himself with increasing vigour. When he wasn't talking to himself, he was talking to hedgehogs. After a night of recording hedgehogs, he went to brush his teeth at the tap in the field. As he exited his caravan, he saw Nigel. Nigel was a hedgehog. Hugh knew it was Nigel, because he knew Nigel's personality.

Nigel scuttled down the country lane and Hugh followed. He watched him, fascinated as he saw him feeding in the verge, picking up small bugs and little slugs. Eventually Hugh decided to lie down in the road next to Nigel, who was not scared. Unlike many creatures in the natural world, hedgehogs do not have a fight-or-flight response. Should a threat arise, their ability to freeze and roll into a spiky ball is off-putting enough for even the hardiest digestive system. Though owls, snakes and weasels might eat a hedgehog, they are lower on the menu than softer, silkier and more tender mammals. Mutation, heredity and natural selection have placed hedgehogs in a position of placid aloofness.*

Nigel looked up at Hugh. Hugh looked at Nigel and felt a change in his relationship with the natural world. When their eyes met, he got the feeling that he was being watched for the first time by Nigel and felt a shift in his connection. Hugh knows the rewards of these connections. While his mind is scientifically rewarded with information from which he can build his ideas and

---

* Similarly, toads are without a fight-or-flight response. Rather than having spikes on their backs, they have secretions. These will soon foam up a dog's mouth and make toads a near-impossible meal.

his enquiry into nature, there is also the rush of endorphins that come from the delight of connection. He has a robin that visits his garden and Hugh has learnt to feed it by hand. When the robin lands on him, he feels the dopamine hit, which generates an entire sensation of feeling good about himself. He sees it helping in its own self-healing way, because he feels better: 'You're not so down, and I'm hoping that for the empathetic amongst us, I can help trigger a small, similar surge of good feeling by doing it. This is entirely scientific, this is not hippie stuff.' He pauses. 'Well, I might be a hippie, but I'm a hippie who is also grounded in science.'

Hugh scrutinizes what he describes as 'the triggering of all those cascades of loveliness', but he never loses sight of it also simply being a lovely thing and a wonderful possibility, which comes from having brains that have evolved in such a way. He knows the point of science is to be utterly objective about testing a hypothesis that is attempting to prove whether your view of the world is right or not, but while you are writing your paper on otter behaviour, you are still allowed to take a break and smell the poo.

During his period of lockdown, Hugh started to obsess about smaller and smaller things, because these were the things he had access to. It's true that, for many people, our frame of focus has become smaller and smaller. When your vista rapidly shrinks, you start to pay attention to the details in the corner of the view that remains. During a period of detachment, alternative connections need to be made. For me, the skylight in my bedroom became the picture frame of the whole available world. I have never been so familiar with Venus or so observant of every dusk.

In the mornings I would exercise in our small strip of garden. I grew close to a daffodil. Through tears of sweat and sighs of resignation, I saw the same trumpet of petals every day. At first the daffodil was in rude health. As the days passed, the stalk lost its rigidity, the stamen dried, the bees lost all interest, moving on to next door's blossoming tree. Eventually bent like a nursery-rhyme crone, the daffodil was dead. I chronicled the daily change on Instagram, and people grew emotionally tied to it until, one day, there were no more pictures. I would have buried it in the garden, but it had already moulded into the earth. No shovel required. The static nature of our existence in the spring of 2020 gave us the perfect time to watch the spring roll over us and pass by.

The celebrated science essayist Stephen Jay Gould wrote that we will not fight for what we do not love, and Hugh very much believes this. It is what infuriates Hugh about social media and its 'like' buttons. Liking is not enough – liking will not get the dishes done or the forests saved. No revolution has ever been born off the back of people liking something, but the problem with opening yourself up to love is that you open yourself up to grief and, for some in a cynical world, the embarrassment of declared love, too. While giving a lecture at the Durrell Institute of Conservation and Ecology, Hugh talked about love. In a room full of academics, he sensed a mix of confusion and disapproval, the sort of reaction you get from scientists if you say something statistically convoluted or declare that there are canals on Mars made by Odin.

Hugh considers himself to be immersed in biophilia, a term first coined by the psychoanalyst Erich Fromm, who wrote of alienation and the human search for authenticity. In *The Anatomy of*

Human Destructiveness, Fromm described biophilia as 'the passionate love of life and of all that is alive'. The biologist Edward O. Wilson expanded on this in his 1984 book *Biophilia*, defining it as 'the innate tendency to focus on life and lifelike tendencies'. Wilson sees it as 'the emotional affiliation of human beings to other living organisms'. The argument is that our encounters with, and contemplation of, nature are not just 'nice' or 'something to do on a Sunday if there is a power cut', but are necessary for our vitality and our happiness.

Writing on perceptual pleasure and the brain in *American Scientist*, neuroscientists Edward Vessel and Irving Biederman have researched how different images lead to different rewards in the brain's opioid receptors. Images of natural landscapes are more rewarding than other images. The movement of waves or leaves on a tree holds our attention for longer periods of time.[10] [11] Representations of nature in art, architecture or ornament have some benefits, but are less effective.[12]

So connecting with nature is not mere frippery; it is something that needs to be widely promoted and made as accessible as possible to all. When my disappointed and fuming teenage son was infuriated by a video game that did not give him his deserved reward, I persuaded him to take a walk with me along the nearby canal. After ten minutes of him being in a tight knot, the sky, the bees and the sight of a heron patrolling a lock all dissipated his screen anxiety and we absorbed the spring scene merrily, waiting for ducklings by a bridge as it snowed blossom. On days like these, I am glad that my poorly illuminated basement flat on a hectic London road flooded with sewage, and persuaded me that the city was no longer for me.

In 2011 Ihab Elzeyadi, founder of the University of Oregon's High Performance Environments Laboratory, published a paper looking at the impacts of access to daylight and the views from windows in offices;[13] 30 per cent of the offices overlooked trees, 31 per cent looked out at the street or a parking lot and 39 per cent were on the interior of the building. According to his research, the view that a worker had access to was the primary predictor of absenteeism: it did not merely affect a worker's happiness, but also a worker's behaviour. Those with views of a landscape took an average of fifty-seven hours sick-leave per year, while those with no view took an average of sixty-eight hours. In a paper on aggression and violence in the inner city, the authors wrote, 'the results of the study found that some types of domestic violence were 25% less prevalent in the greener housing developments compared to the barren housing cluster'.[14] Obviously, like everything in research, it is, more often than not, rather more complex than that, but the desire for nature, the effect of our connection with other living things and its importance are neither easy to dismiss instinctually or from scientific research.

'The most wonderful world that ever worlded'

I spoke to the writer Neil Gaiman during a period of turmoil for him. He was on the other side of the world from his family, trapped by the Covid lockdown, isolated on the Isle of Skye. He was heartbroken to be alone, heartbroken to be separated by oceans from those he loved, but he had found solace in seeing the glorious foxgloves and looking at the bluebells. This was the flipside of the cold and pitiless universe; looking at the seals in the sea and the cloud-veiled mountains, he sees 'the most wonderful world that has ever worlded'. A few days before, he had gone out on the night of the Summer Solstice and looked at the most wonderful sunset. His heart was healed. He asks, 'What makes it worthwhile? Why are we here? And you look at that sunset and you go: This is amazing.' He walked up the hills and found himself remembering some incredible knowledge that he had filed away and forgotten, and it blew his mind. He recalled that the rocks of Skye are the oldest exposed rocks we have on this planet, apart from the Himalayas – the 3,000-million-year-old rocks of the Lewisian gneiss. He could not be sure it was true, but it was in his mind that moment and he felt incredibly connected to the ancientness of it; and sometimes you don't have to go straight to Google to check whether your transcendent moment is factually justified.

Studies such as 'Soundscapes of Wellbeing', a collaboration between the University of Exeter and the BBC, have shown that even the smallest connection to nature will mean that people heal better, heal faster and are less likely to be ill for as long. With the lockdowns of the Covid-19 pandemic, 'Soundscapes of Wellbeing' was driven by the need to find out if there are therapeutic ways of connecting with nature when nature is not physically accessible. The experiments will see if digital communion with nature, such as looking at rich visual scenes or listening to immersive soundtracks of nature, will have positive effects.

For those of us fortunate enough to be brought up in the countryside, it is easy to fall into ambivalence towards the natural world. I had a father who was a keen amateur wildlife photographer, so I grew up trailing behind him across hills and meadows as he tried to capture all the butterflies of the British Isles on film.

I can't quite recall if we ever managed to sight that elusive marsh fritillary that was the focus of a summer holiday in Dorset. It also meant that I was pushed into rivers when he lost his balance, then tried to use me as a fixed post to steady himself. My father would remain dry and I caught a cold, but isn't that what sons are for? I also witnessed the gutting of rabbits and the hanging of dead birds. I mainly eat tofu now.

It is only in middle age that I am truly realizing the importance of those treks and sights, which were so much more than wet boots, the pain of nettle stings and ladybird distractions, though I am still not sure I will ever come to terms with how brilliantly preposterous the lifecycle of the butterfly is. When looking at the beauty of its wings, and the seeming absurdity of the journey from caterpillar to cocooned pulp, to butterfly, it can seem wonderfully absurd; but, like the quantum behaviour of a particle, we have to remember that it is absurd to us, but quite usual for the laws of the universe. As Edward O. Wilson wrote, 'Life around us exceeds in complexity and beauty anything else humanity is ever likely to encounter.'[15]

I don't think life has become reduced to letters, now that the codes of our construction are available to us, but instead life becomes more beautiful, now that we have this code to unravel and translate us. Our closeness to other living things has become measurable – another occasion where information creates connection. We can go back in time to work out where we join hands, paws, tentacles or gelatinous fronds with other species, and the shared ancestors on our family tree.

Professor Aoife McLysaght is a geneticist currently working at the Molecular Evolution Laboratory of the Smurfit Institute of Genetics in Dublin. She is also a passionate and witty public speaker, and presented the Royal Institution Christmas Lectures with Alice Roberts in 2018.

Aoife's path to genetics came from an inspirational biology teacher who loved teaching and went way beyond the requirements of the curriculum. He didn't explain everything. Sometimes he would just leave clues to theories and ideas, which were beguiling to the more curious student. During the term he spent on genetics, there was one particular three-letter clue to further knowledge that Aoife noticed on the overhead projector. Whenever he was writing a gene, he would always write ATG and then a squiggle. He didn't write any of the other possible three-letter combinations, of which there are sixty-four. This ATG and squiggle made Aoife wonder and puzzle. Later she would learn that ATG is a very specific three-letter combination in the genetic code, because it's the start code.

From the moment evolution came into her education, it made total sense to Aoife, but it wasn't until she was giving a lecture about incest and bestiality at the Electric Picnic Music and Arts Festival – two subjects that are more likely to be the focus of a stand-up set than a biology lecture – that she had a revelation about connections through multiple strands of DNA, which has since given her a real sense of joy in life's magical meaning. She realized, whilst taking time out thinking at the festival, that every piece of DNA has a physical connection with every other piece of DNA. Her DNA hasn't touched my DNA, but they have touched, in a sense, in our ancestors. When it came to her, it seemed blindingly obvious, but also revelatory. 'I remember when that occurred to me, which is quite recently, considering I've been studying genetics for such a long time. I just had this moment of going: Wow!'

The physical touch comes from the necessity of DNA unwinding. Each half becomes a template for the building of the new half. Like Apollo 9's Rusty Schweickart, a revelatory flash has been in your Aoife's head for a long time, gestating, waiting for the moment when it all comes together. Here was the unravelling connective tissue spread like silky webs across the branches of our family tree. None of us have met our great-great-great-grandparents. We know that we're related to them, but we rarely think there is a physical connection. Sometimes it might strike us when looking at a very old family photograph: we see the nose on the face of an ancestor and realise that the code of that nose is the code of our similarly aquiline or bulbous nose. It is an idea of life that demolishes boundaries: with all these people we have never met and, expanding from that, every living thing on this planet, there is a little chain that links it.

Aoife still berates herself for taking so long to come to this realization, when she believes it is the natural conclusion, from things that had been familiar everyday knowledge to her. There is a beauty in this reminder that even a scientist can still be startled by a world they have been immersed in for a long time. Just as you might find your keys when you stop looking for them, so you might find a mind-enhancing vision of the physical sense of the connective tissue and code that join together the entire living world, right as you were wondering where you were going to get your first pint of hot cider, after discussing incestuous bestiality in a field, near Radiohead.

Aoife wanted to tell everyone, 'It's not just that we're related. It's that everything has touched everything – maybe at a certain number of points of removal, but everything has touched everything. Everybody's touched everybody, and everything and every living thing.' This is of course not what a keen racist wants to hear. Not only do we have hankerings for human exceptionalism, but some people demand that certain groups of humans are more exceptional than others.

As soon as Aoife starts trying to talk about any genetic difference between different groups of people – and especially when talking at all about genetics and intelligence (and it does appear there are genetic factors that influence intelligence) – then it attracts racists who say, 'And white people have all the good ones, right?' Those hankering for their own biologically innate superiority are keen to find hard and clear lines that are clearly attached to a very specific colour chart. Aoife thinks deliberate misinterpretation is almost impossible to avoid, because 'People will contort themselves sufficiently to come up with the conclusion they had already, but then by bringing it up, maybe you can knock down some of that misinterpretation at the same time.'

She explains, though, that if you take any two people from anywhere in the world and ask how many differences there are in their genome, it is roughly the same number. If you take you and somebody that you are not related to – an unrelated person from Britain, France, Nigeria or China – and you count how many differences there are in the genome, it is always approximately the same number. I think this a very powerful piece of persuasive information that helps explain that we are all the same, even in our differences.

Making sex boring

Adam Rutherford, geneticist and author of *How to Argue with a Racist*, also finds the biological tree of life and the superficiality of human differences genuinely warming. His frustration lies around the difficulty of this idea being universally accepted, and how if it could be embraced, then we might spend less time killing each other or contemplating the vanities of our small differences. Adam says he revels in the complexity that biology gives us, and that understanding of nature being a total organism.

He doesn't believe that his understanding of evolutionary biology and life on Earth enriches his personal sense of meaning, purpose or place in the universe, although he thinks a lot about duty and legacy. Adam says, 'If you accept the notion we have eighty years, that gives you a sense of responsibility to fill it with joy and work and value, and all the basic humanistic values' – a position that he has focused on even more, having recovered from a serious bout of Covid-19. Understanding life on Earth is one of the ways in which he has tried to achieve those goals. In terms of legacy, Adam is aware that can seem like a vain ambition, and his hope is that the legacy is most clearly evident in the behaviour of our children: that they might have self-reliance, joy and behave better than us.

His inspiration was his genetics tutor, Steve Jones, who has summed up biology as 'Life's a mess' – and therein lies the fascination and frustration. If you want demystification, you might want to go into physics instead. Physicists may say, 'The more you know about physics, the simpler physics becomes.' You get universalist models of physics that work, but the more you know about biology, the more of a mess it becomes. Steve Jones is a geneticist, science writer and snail expert. Whatever the question, he will normally find a way to get snails into the answer. He has a remarkable ability to pick up snail shells without you noticing. Go on a country ramble with him and, by the end of it, his anorak pockets are bursting to the seams with shells. He is a gastropod prestidigitator. There are more genetic differences between two snails in different valleys in the Pyrenees than there are between chimpanzees and humans. Steve says that his job as a geneticist is to make sex boring, but the reward is that he makes snails very interesting.

Steve explains natural selection in a captivating, slightly grouchy manner. He describes the widespread use of the peppered moth as the central example to explain natural selection as 'the world's most boring, boring example'.* Steve feels this common example fails to give a sense of the power of natural selection, which he sees as a factory that makes almost impossible things possible. Natural selection is a series of successful mistakes. This is the uplifting idea of mutation, heredity and natural selection over manufacture by a deity. You are not perfect. You are not meant to be perfect. You are a mistake, but you are a better mistake than those mistakes that didn't make it. You are the least-wrong shape and form, and you come from a long line of people who were less wrong than the competition that didn't make it to here.

---

* The peppered moth had a pale colouring suitable to camouflage it on the bark of trees, but during the rise of industrialization the moths become dominantly black, enabling them to blend in with the soot and grime-covered landscape. In 2016 Dr Ilik Saccheri and his team at the University of Liverpool discovered the specific mutation that leads to 'industrial melanism'; https://news.liverpool.ac.uk/2016/06/01/jumping-gene-took-peppered-moths-dark-side/

The point about biology is that 'it is flexible and it is constantly changing'. Before Darwin, species were fixed. There was no intricate tale of the evolution of a bird's tail or beak or wing. Darwin enjoyed collecting beetles, searching for those beetles that had not yet been observed and catalogued. He was a naturalist; he had no real interest in why there were forty species of orchid in this meadow – they were just there and he ticked them off. This picture began to change during the voyage of the *Beagle*, as Darwin found that he was drawn to the cumulative peculiarities of the distribution of the animals and plants of South America.

When Darwin observed the animals and plants on the Galapagos and Cape Verde Islands, he started to think they should be very similar, as they were volcanic islands and surely a designer would use the same design for the same terrain; but the living things on the Galapagos Islands looked more like living things on the nearby South American mainland, and those on the Cape Verde Islands looked more like things on the nearby African mainland – and, from there, the rich story of adaptation grew. For Steve, *On the Origin of Species* is 'a masterly piece of propaganda. It starts off with something which is bleedingly obvious, about milk and dairy cattle. He [Darwin] points out that the cattle are very similar to each other, but they've changed a bit. He does it with other domestic animals as well.' From there, Darwin builds an argument that concludes with 'light will be thrown on the origin of man and his history'.

Now there is a cliffhanger, but the reading public needed to wait twelve years for *The Descent of Man* to provide the answer. *On The Origin of Species* is a beautiful story that enriches any living thing you can look at. Life is no longer a designed product.

Instead its shape and form have a story, and in that story the shapes and forms of everything around it – from landscape, to predators and climate – are all part of it, too. Each species is interacting; there is no space for narcissism in the behaviour of a bee, a beetle or a birch tree. Just as the story of the universe becomes so much more dynamic when it has no longer 'always been there', so the dimensions of life become far more beguiling and intriguing. There are so many more adventures to behold.

After all these years of Steve looking at the theory of evolution and observing the changing tools of understanding, I wondered if it had changed the perception of himself. Steve doesn't think so. For him, it acts as a memento mori. 'Once you accept the facts of death, that's it – you're in the hands of biology and there's not much you can do about it.'

Thanks to the little I have learnt about biology and its revelations, the flowers no longer look the same to me. The beetles or bees that I see in them have changed them, as have my thoughts on the colours of their petals. The beauty of the flowers is not a full stop; it is a question mark. Why does that flower grow here; why is that meadow filled with those shrubs; and what has brought these tiny insects to crawl all over these dandelions? I stop at tangled banks more often since I read Darwin: 'It is interesting to contemplate a tangled bank, clothed with many plants of many kinds, with birds singing on the bushes, with various insects flitting about, and with worms crawling through the damp earth, and to reflect that these elaborately constructed forms, so different from each other, and dependent upon each other in so complex a manner, have all been produced by laws acting around us.'[16]

I also think of the relief in knowing that perfection is an impossibility, but that we should not beat ourselves up when we see our own imperfections, as imperfections are all part of the process. Life is mistakes.

Brewery contemplations

A lecturer on evolution was once interrupted by a concerned student who asked, 'Are you really saying that we are all related to monkeys?'

'Better than that,' replied the professor. 'I'm telling you that you are all related to yeast.' Since that day, our connections with yeast and all other living things have become even better known.

Paul Nurse and Peter Higgs are the only two Nobel Prize-winners I have ever had to explain anything to – I'm keeping a tally, which I don't think is likely to increase much. With theoretical physicist Peter Higgs, I explained how a specific note from a cello could stop a crocodile from attacking you, although the distance that the Royal Philharmonic is from most lagoons means it is unlikely to be anything more than an intellectual exercise – and it is nothing that I have put to the test, as cellos are hard to drag through undergrowth. Harmonicas are apparently a less effective defence against crocodiles according to a one-armed man I met in a bar.

With geneticist and former President of the Royal Society, Paul Nurse, I had to explain how a car door worked – specifically the very idiosyncratic handle on a Birmingham taxi. I'm not ashamed to say that there are unlikely to be other times in my life when my qualifications surpass Paul's. Among the very many subjects that he is better qualified to talk about than me is yeast. What he discovered about yeast, with the vital assistance of Melanie Lee, whom he cited in his Nobel Prize acceptance speech, is a story that I think makes life even more beautiful and astounding.

Paul's interest in what life is began when he saw a yellow brimstone butterfly. His butterfly fluttered, then settled. Paul disturbed it, and it went over the garden fence. It was then that he began to think, 'What is it that is similar between me and other living things like that butterfly?' They were obviously very different, but also very similar in many ways. This seemed to be a fundamental question of biology. What is life? How do you find it? How do you distinguish between something that's living and something that is not living? What does all life share?

Thanks to Paul's work, it seems that one thing all life seems to share is cyclin-dependent kinase 1, something that plays the key role in cell-cycle regulation. In humans it is encoded by the CDC2 gene. When you delight in the smell of freshly baked bread, think of the connections that run through all life on Earth. As a research scientist, Paul has an obsessive interest in yeast and how it works. The particular yeast that he focuses on is called 'fission yeast'. Though you may not be fully conversant about nuclear division in eukaryotic cells within a granary loaf,* as you spread your peanut butter on a slice, you can dwell on the fact that the lifecycle that led to that and the cycle that led to you share some stories. As Paul observed the reproduction of cells in yeast, the question arose of whether the reproduction of human cells was controlled in the same way?

---

* If I'm honest, you won't have much of a conversation with me about nuclear division in eukaryotic cells, either, so please don't think I am trying to show off my superiority to bread.

At the time Paul could understand why this might be seen as a totally absurd question to ask. Aren't we so much more complicated and magnificent than yeast? After all, yeast is one of the simplest eukaryotic cells* in existence – and we are able to ride bicycles, write poetry and do traditional country dancing, should we wish to. To test the hypothesis, he struggled with conventional methods and failed. At a point of desperation, he suggested an experiment in which they would use a human gene library, something that at the time had never been done before. He describes the process as 'basically sprinkling this human gene library onto a particular yeast strain'.

He reasoned that if there was a similar gene to that in humans, and if they were to sprinkle that onto the yeast cells, and they took it up and it worked perfectly, the cells would be able to go on and reproduce. They would form a little colony on the plate. Paul concludes his description of the experiment with, 'Isn't that absurd? Is that likely to work? There's about fifty-two "ifs" in that reasoning there.'

They did it, and they got colonies growing. Paul thought it couldn't be right, imagining that there must be a contaminant – just like Alexander Fleming's bacteria flying in through the window. They remained in suspense for a couple of months. He knew that if it worked, this was big. They ended up sequencing the gene. Paul saw it come out of the computer like ticker tape. The letters came up. What it predicted was that the human gene was 61 per cent identical to the yeast gene. It was 297 amino acids long, and the yeast gene was 298 amino acids long. This

---

* Eukaryotic cells are ones that have a nucleus.

was not the gene itself, but the protein that the gene makes. The same process that worked for the human gene worked in yeast.

Paul explains that this simple experiment tells us some very interesting things about life. Firstly, that the way in which cell reproduction, which is one of the basic properties of life, works in yeast – one of the simplest eukaryotic cells that you can find – is exactly the same as in humans, to the extent that you can take the human gene and put it in there and it substitutes. Secondly, it means that everything between yeast and humans, which includes fungi, plants, insects and apes and anything living in between, is extremely likely to be controlled in exactly the same way. This turned out to be true. Yeast and humans diverged from one another on the evolutionary tree probably around 1.5 billion years ago. It means that this method of cell reproduction has been conserved all that time and it still works.

Paul's first reaction to this was complete wonder. This then extended to a contemplation of what it meant philosophically, as well as its biological implications. It means that we are fundamentally related to every living thing on this planet. Paul sees the philosophical outcome as being that we should look upon all life (possibly with the exception of bacteria) as our relatives. It may be a long time since we shared a common ancestor, but all living things are our relatives, and should we not feel some responsibility for our relatives? This gives us yet another reason to respect and care for the biosphere, because everything is related to us. We need to respect other life forms not only because we all interact with each other, but also because we are completely dependent upon other life forms in our biological past for our very existence.

I began this chapter by talking about Creationists. Why should it matter that people hold such beliefs? Denying evolution denies our connection with the natural world and robs us of a far richer story than any of the alternatives that I have heard. I think it also robs us of responsibility, which comes from connection and a realization of our proximity to all other living things. Pondering your connection with any living thing starts as an entertaining exercise and can lead to little explosions of revelation. You might notice someone sitting opposite you on a train and note that they have facial features and colouring close to yours, and you can ponder how many decades you would need to travel back to find a common uncle on your individual family trees. You might look at a bonobo ape showing off with a dead rat worn on in its head as a fashion accessory and, as you adjust your own beret, think that it was six million years ago that events and environments saw your common ancestor branch off – one to evolve into that bonobo and the other to evolve into you.

Look into your bowl of vegetable soup (I am presuming you have given up pea-and-ham since you watched that clip of Pigcasso) and realize that if you go back far enough, that carrot and you were on the same family branch, before circumstances started the separation that led to your situation on the chair and the carrot's situation in the bowl.

Sit up a tree with an orang-utan that, just fourteen million years ago, you had that falling-out with, and you both walked away from your shared ancestor; or maybe shake hands with the axolotl that you branched away from only 350 million years ago. To me, these connections offer so much more than being a singular creature, aloof and utterly separate from all the living

235

beauty. Our connections elevate us and enable us to gain a deep understanding of our history and how we came to be as we are.

> The more clearly we can focus our attention on the wonders and realities of the universe about us, the less taste we shall have for destruction.

RACHEL CARSON

CHAPTER 8

The Mind Is a Chaos of Delight – On the Matter of Grey Matter

Insight roams the sea of the unconscious like the Loch
Ness monster, a rumor whose wake occasionally becomes
visible, but even then it's mystifying and scarcely believed.

Diane Ackerman

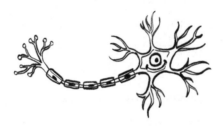

I am going to preface this chapter with a warning. It may be the
most unruly of all the chapters in this book, and the reason for
this unruliness is the very subject I will be talking about. When
I focus on my brain, my brain becomes particularly feverish and
excitable. It's as if it is saying, 'You're talking about me now –
tell them this; no hang on, I've got another idea, what about
this? Oh, and don't leave this thought out.' Thinking about the
brain – the citadel of all our experience – leads me to a cascade of
thoughts, all talking over each other. Rather than a citadel, it can
seem like an old bombsite, rooted over by scavengers throwing up

any bit of rubbish they find, then running to another decimated bicycle or shattered vase.

I almost skipped the whole subject, but one night in a dressing room I heard a comedian feeling anxious about his brain, and even more suspicious of neuroscience's quest to understand it. He saw it as a breach of his rights that scientists might one day know so much about our brains that they could see our thoughts and make educated propositions about our actions. He had a conspiracy mindset, when it came to his mind. It was a dystopian view of neural enlightenment and, from my perspective, wrong-headed – although I will only truly know if he is wrong-headed when my neuroscientist friends finally finish making that machine I've asked them to construct, so that I can look inside his head while he sleeps.

Should you find yourself confused by this jabbering over the next few pages, consider it part of the explanation of the strangeness of our mind. With this caveat, let's talk about our brains.

Psychology researcher Julian Jaynes wrote that our brain is 'an invisible mansion of all moods, musings, and mysteries, an infinite resort of disappointments and discoveries. A whole kingdom where each of us reigns reclusively alone...'[1] Poet and author Diane Ackerman described our brain as 'that mouse-gray parliament of cells, that dream factory, that petit tyrant inside a ball of bone...'[2] Reading these elegant descriptions might make our hearts swell with pride as we think of the billions of years of mutation, heredity and natural selection that have gone into creating something so elaborate and intricate – and that it is ours. At least, it seems to be ours.

There have been some unsettling revelations in the last few decades about our brain: this fabulous mass of living tissue that controls our perception, our actions, and contains our self and our identity. In the last hundred years our free will and our selfhood have not only been brought into question, but have reached the point where some researchers have dismissed their existence altogether. You are a delusion, and the idea of controlling your life and actions is merely an illusion. Neuroscience can appear to rob you of your autonomy, your personality and you; and, on top of that, the seat of your soul is a matter of matter alone. Such ideas can be terrifying, but also liberating. After spending years of your life self-flagellating over your errors, it seems you might have been a passenger all along.

Getting to know your brain is not always comforting, but it is very worthwhile; and, at least until the end of this book, I will remain deluded enough to believe that I – the me-ness of me – exist.

One bastion of cutting-edge brain research is *The Simpsons*. Homer's brain has often been at the forefront of neuroanalysis. His brain has had vibrant conversations with itself, and it has found Homer so disappointing that it has even walked out on him. Episodes have frequently highlighted notions that we are many personalities in our minds, often at loggerheads, all battling with each other and eventually being funnelled into one person. In the episode 'We're on the Road to D'ohwhere', Serious Homer Brain holds a gun to Fun Homer Brain, when the latter wants to get a monkey drunk and pushes it down the stairs. Serious Homer Brain also reveals that he murdered Intellectual Homer Brain.

In the 'Mr Plow' episode, after Homer crashes his car, an insurance salesman asks him where he has been. After remembering that he has been at Moe's, Homer's brain tells Homer that he must make sure he doesn't reveal that Moe's is a drinking tavern. To avoid this embarrassment, the alternative he comes up with is, 'It's a pornography shop. I was buying pornography.' Homer's brain proudly looks on and says, 'I would never have thought of that.' Here we watch a process that we have all had in a panic, or when seeking an alibi: all those voices seeking the best answer, and the pride on the occasions when it actually finds its way out of our mouth.

Homer's brain has been imagined as a sleeping fly-covered horse, a clockwork, cymbal-playing monkey and a series of simplistic, ancient monochrome cartoons. Homer's brain is many things and many people. A cartoon maybe, but this all fits with neuroscientist David Eagleman's belief that we are made up of 'a team of rivals'.[3]

Most of us have a superior brain that, annoyingly, always seems to make itself known after the event. It's slow to react, but it is smart. It leaves humdrum brain in charge, but always turns up in time to explain what it would have done, had it been there at the time, and how it would have dealt much better with the situation than you did. Some days it can seem like a machine that generates perpetual regret, guilt, shame and self-loathing. Worryingly, this may reflect a scientific reality. Neuroscientific research suggests that we live our life in hindsight. We unconsciously react to the world, then explain why we did so afterwards. *You* always only ever turn up after the event. Reassuringly, it might mean that you don't have to spend quite so long blaming yourself for every mistake and clumsy gesture.

A sense of responsibility, a sense of being in charge – that circle of coaches and critics talking in your head – may be a peculiarity to our species. As I am writing this, there is a particularly annoying fly buzzing loudly like a lisping kazoo. It is a lazy late-autumn fly, plump and monochrome. It combines an endless drone with clumsy collisions into windowpanes. It can be hard to believe that all this annoyance, which I am beginning to take so personally, is totally without intention.

On my more misanthropic days, I can feel like this about other humans. When stuck on a heavily delayed train, next to a loud chewer with a persistent cough and not the slightest intention of covering their phlegmy mouth (there are some things I didn't miss in Covid lockdown), I wonder if some of us are so lacking in self-awareness that we, too, may have all the consciousness of a bluebottle or a gnat. Despite this, I still hold back from the killing of the lazy fly (or the consumptive commuter). Should I swat it, I would experience a wave of guilt that this insect would be totally incapable of, if it killed me by poisoning my marmalade with its vomit. All evidence so far predicts that bluebottles do not spend their life looking back and wishing they had realized that going so quickly from prodding dog excrement to landing on jam may have terrible wider ramifications. Flies are not known for their regrets, but then again, neither are politicians, at least when it matters.

The search for self-consciousness, for the neural heart of us, can seem esoteric, a hotly debated whimsy, but the stops on the journey can be life-changing for some.

'If we don't do science, these problems will persist!'

Our level of self-awareness is something neuroscientist Adrian Owen has been thinking about for a long time. When he was training to be a neuroscientist, he went camping with another student. One night they sat on the branch of a tree and started to wonder if some people were more conscious than others. Some people seemed to coast through life, not noticing much, their minds on that standby setting, like a driver taking a regular route, where awareness only kicks in should a deer leap into the road or a horde of zombies start banging on the rear window. Others spend their life with different levels of hyper-vigilance, always envisaging a rabid goat around the corner or expecting the shadows to become phantoms.

I am in the latter group. Every time I walk down the street, my mind is highly likely to wander into working out the best escape route from the undead hordes. My mind is often in the pre-title sequence of a movie that doesn't turn out well for anyone. I rarely mentally switch off. Adrian's work now deals with people who have often been dismissed as being switched off. His work is about making contact. He deals with a very pragmatic understanding of consciousness, and believes it is study that can transform someone's life.

I first saw Adrian when he was giving a lecture about detecting consciousness in patients who appear to be in a vegetative state – this is different from being in a coma or locked-in syndrome. Someone with locked-in syndrome is not cognitively impaired, but has lost the ability to move their body. A coma has the appearance of sleep. Someone in a vegetative state, though, has sleeping and waking cycles and may seem to look around, but at nothing in particular, and makes noises such as snorts, though these are not attached to exterior stimuli.

It was generally believed that someone in a vegetative state has no conscious awareness. Adrian's first case was Kate, who was placed in a scanner – something that was considered pretty pointless, due to the cost and the belief that she lacked any awareness. It was believed that *she* was not there any more; her breathing body had no self left in it. In the fMRI scanner, it was discovered that if Kate was shown a photograph of someone she knew, the facial-recognition area of her brain lit up. There were controls. If a blurred photo or a stranger was shown, then the recognition reaction did not occur.

Kate now describes this moment as being the point when she went from being a body to being a person again. A scan found 'her'. The area of the brain that lit up is known as V1. If scrambled images were shown to Kate, the area did not light up, as the reaction required active recognition. Kate is now able to communicate and lives a full life. She sent Adrian this email:

Dear Adrian,

Please use my case to show people how important the scans are. I want more people to know about them. I am a big fan of them now. I was unresponsive and looked hopeless, but the scan showed people I was there.

It was like magic, it found me.

Love from Kate

When we see someone who appears to be bodily intact, it is hard for us to believe they are not accessible. It is hard to believe they can't be 'woken up'. Such considerations were what provoked the battle over Terri Schiavo's right to live or die. Terri was in a persistent vegetative state after being starved of oxygen, due to cardiac arrest. Was Terri still inside her body? Her husband fought for her right to die, as he did not believe she would have wished to have a prolonged existence on life support. This was a battle between family members, and it became a political issue stirred up by the religious right in America.

Adrian explained to me that if he was to use anyone in a vegetative state for campaigning against the right to die, then Terri Schiavo was not a good example. When Kate was put into an fMRI scanner, her brain looked normal. Her external behaviour may have been abnormal, but her brain was not. When Terri Schiavo was put into a scanner, large bits of her brain were missing, because loss of oxygen to the brain had meant that cells in the brain had died, and large bits of her brain simply weren't there. Adrian believes that consciousness is a physical property – a physical outcome of having a brain as sophisticated as we have. If you don't have a brain, you can't be conscious. If there are bits of your brain missing, there are things that are going to affect your consciousness. He does not believe there was any possibility Terri Schiavo had any awareness, because she didn't have any of the neural architecture that you would need to be conscious, whereas Kate did.

Adrian's research leads to questions of where our consciousness is, and what we should expect from it. His approach to consciousness is pragmatic. He explains, 'I'm not gonna waste a lot of time trying to find it. I come from a starting point that I am conscious. And what I do is I investigate other things – be it vegetative patients, chimpanzees, babies – and I try and work out "is their inner world like mine?"' If he decides that they are like his, then he's perfectly willing to say they are conscious. He doesn't need to define it any further than that, or get caught up in philosophical arguments. He is measuring properties. He wants to try to solve problems for patients.

Sometimes you need Alfred Hitchcock to help find out how the brain works. Next time you are watching *North by Northwest* as Cary Grant is being chased by a crop duster, or Janet Leigh is stepping into the shower in *Psycho*, take a moment to wonder what your brain is up to. It may well have been taken hostage. Films hijack people's consciousness, at least good films do. Adrian describes it as 'surrendering your consciousness'. You become unaware of your own surroundings and instead you are in the story. For Adrian, this is the best sort of experimental model.

He finds that films like the *Fast & Furious* series are not as useful as a Hitchcock thriller, as there is not a singular element that everyone is attached to – some people might be ogling the cars, some people are mainly drawn to Dwayne 'The Rock' Johnson, other people might be making vrooming noises while imagining they are Vin Diesel. And so films of multiple characters, high energy and noise distractions are not the first choice for the neuroscientist at work. Violent action films create a different internal scenario, as things like guns going off light up the auditory cortex, and that is not a consciousness thing.

Adrian's preferred film in his studies is a thirty-minute episode of *Alfred Hitchcock Presents* entitled 'Bang! You're Dead'. It is the story of a child who finds a gun that he thinks is a toy, but which the viewer knows is a loaded weapon. The plot is simple and suspenseful. Adrian has observed viewers' brains in a scanner and found that our reactions are synchronized to the action onscreen. This is the ultimate form of film criticism – a truly engrossing piece of art will see the pattern of the brain illuminating in sync. If it were possible to scan a whole cinema screening at once, it would be a rhythmic throb of coordinated lighting for the running time of the whole film. No longer do we need film critics to write reviews; they can simply have their brains scanned at every screening and the thumbs-up/thumbs-down review will be replaced by a couple of brain shots and the dopamine count.

If thoughts are just matter, do they matter?

Yet if our brains are merely made up of electrically firing atoms, then how are they able to contemplate other atoms? Our inability to understand what creates this level of self-awareness leaves room for some people to fill the apparent void with a soul, or something ethereal. Adrian believes it is inevitable that one day we're going to work out how consciousness works and where it comes from, but that it remains a complicated problem, and we should not to expect the answers to it any time soon. Those studying the human brain find comfort in the notion that if the brain was simple enough for us to understand, it would be so simple that we wouldn't be asking the question in the first place.

Adrian believes that both chimpanzees and toddlers are conscious, but he also firmly believes that their conscious experience is not the same as his. He has learnt mostly to ignore behaviour when he is trying to work out if something is conscious. As he has seen with vegetative patients, if people had judged those in a vegetative state based solely on behaviour, 'We would still

246

be assuming that all of these patients were unconscious, because they don't behave – they don't do anything.'

I wonder where consciousness goes at certain times. It is costly, in terms of energy, and it is tiring always to be aware. As someone who doesn't switch off very often, I am jittery, twitchy, usually pacing and anxiously curious, so I am fascinated about what happens when we do seem to disappear. A friend of mine was at the beach with his family one day. His daughter was happily out swimming in the calm waters. Seemingly out of nowhere, a tidal swell engulfed his daughter. He ran into the sea, swam out, grabbed her, got her back to the shore. At least that is what he presumed must have happened, but he has no memory of doing any of that. Where was *he* then? Did a biological imperative usurp consciousness? Just when are *we* really required in day-to-day life?

If Adrian could, he sometimes thinks he would downgrade his consciousness to that of a cat. He has three cats and they seem to have a blast. They don't need to worry. They eat, they lounge, they sleep: that's about it. Adrian thinks that cats have some elements of consciousness – enough of a mind to appear to know how to woo a human into buying salmon chunks in a rich jelly – but not so much that they ponder whether taunting a half-dead sparrow in a fit of pique is ethically going too far.

The pottery of your brain

By understanding what the brain consists of, and our mind within it, we can really illuminate why we are who we are, and find new ways of dealing with some of the more troubling and damaged parts of us, if necessary.

Heather Berlin is a cognitive neuroscientist whose particular interest lies in the neural basis of impulsive and compulsive behaviour. She has young children and, in the tradition of Charles Darwin and the psychologist B. F. Skinner, she finds them interesting research subjects, whilst making sure that all her study is done meticulously, and within the boundaries of guidelines and ethics committees.

Her opinion of the consciousness of a baby is that the parts of the brain that put structure on things have not fully developed yet, so it is at this point of life that we get closest to pure sensation. Her view is that, like a psychedelic trip, a baby experiences the world as pure experience, without the overlay of the prefrontal cortex of the undeveloped brain. That would give us structure and would organize all the information. 'It's the frame that creates a perception, but really it's just information coming in, and then being organized in such a way where a perception emerges.'

Neuroscientist David Eagleman explains it as: 'You've got about 19,000 genes, which set things up in an extremely sophisticated way, but from there, Mother Nature's great trick was to drop a human brain into the world, essentially half-baked. And then we allow our environments to wire up the rest of it.'

So much of what shapes who we become occurs in our early years, and yet those events will be unremembered. We do not appear to be there when we are becoming ourselves. We are at the time that we are becoming, but we are not quite there, so these events do not happen to us, but they make us *us*. Heather sees this difficulty. 'There are a bunch of forces and experiences that have made me who I am and I probably don't have conscious access to most of them.'

For those who have experienced early childhood trauma, there can be a deep feeling of 'Okay, that's that. There is no escape from the damage done.' Heather accepts that this can disadvantage people and make them feel stuck, but from her research she believes that there is no reason why you can't reformulate some of those early life experiences and make new connections. There are critical periods of development when you are growing up when the indentations of trauma are greater, when the clay of us is more malleable, which, as we age, solidify, but they are not so solid that we can't change; it just becomes more difficult.

Heather also explains that in your hippocampus – the memory centre of the brain – and in the olfactory areas there are stem cells that can migrate out at any age and develop new neurons, making new connections. It is harder in someone who is seventy than it is for someone who is five, but it is not impossible.

Heather had a seventy-year-old patient who had been depressed his whole life. He didn't want to die never having experienced happiness. She worked with him intensely. He had been in therapy for many years of his life, but he ended up trialling the horse tranquillizer and nightclub drug ketamine. He had intravenous ketamine treatments in conjunction with therapy. In what might be considered the resting state of the brain – that period when you daydream or doodle thoughtlessly – depressed people can have their mind hijacked by negative thinking and rumination. Ketamine can release the patient from this default mode by hijacking the negative thoughts. When released from rumination, other things can come in and you can remould in that resting state. Ketamine opened up the patient's brain to allow for new connections to be made. He found a way of feeling happiness.

Some of us hanker for the blank-slate model of the child brain, while others might hanker for it *all* to be determined by our genome. Heather thinks personality as a whole is genetic and she explains, 'It's fundamental to who you are. From your temperament as a child, we can predict your personality traits all the way up through adulthood. So personality remains fairly constant, but the way you're thinking and those patterns of thought can be modified.'

She has had a lot of discussions and debates with developmental psychologists and attachment theorists about the importance, or otherwise, of having a healthy environment when growing up. Whilst Heather believes that attachment is vitally important, she has heard some people say that unless you have a positive attachment figure in the first few years of life, you'll *never* be able to form healthy relationships. She considers this to be wrong and relates it to people who come back from combat with post-traumatic stress disorder (PTSD). Each time they relive the traumatic experience, they strengthen the connection between the events and the negative emotions.

The aim with a traumatic event is to be able to bring it to mind, in a neutral or even a positive environment, and either dissociate it from the emotions connected to it or form a new connection with a neutral or positive emotion. The negative emotions are often running in the background like a programme, triggered by anxiety, fear or panic, but under the surface. Such trauma needs to be brought to the surface and then reintegrated in a new way. MDMA – or ecstasy, as it is known if you are dancing and blowing a luminous whistle – is now being used with some PTSD patients in therapy. Until this point, my general experience of MDMA came from music festivals, where those on it lost all sense of personal space and prodded me again and again with their overactive glow-sticks.

In therapy, though, while the patient is on MDMA, negative emotions are brought up and reintegrated in the brain in a more neutral or even positive way. The negative emotions are no longer running in the background, causing the anxiety, stress or terror. Heather has seen it work surprisingly well. She has been involved with trials involving MDMA and PTSD, where nothing else has worked; where the patients have tried every SSRI* and are still having suicidal thoughts. MDMA, though, seems to work.

Heather thinks that if we understand the brain as chemical and biological matter that is not imbued with an ultimately unknowable mystical property, then it gives us a far greater opportunity to learn how to change it for the better. An ethereal soul, on the other hand, is a lot harder to nail down or pep up. She remains very enthusiastic about psychedelic medicine. She doesn't necessarily think it is the neurochemical effects that are important, but the psychic effect: the release of certain constraints. The prefrontal cortex suppresses things and keeps them at bay because that is effective, but when we need to release that suppression, in order to allow those unconscious thoughts to come to the surface and reintegrate them, the psychedelics can step in and enable the lifting of the constructs. Drugs that were once demonized may now offer us hope.

---

* There are a lot of acronyms in psychiatry – SSRI stands for selective serotonin reuptake inhibitor, which helps to increase serotonin levels in the brain.

Are there still teenage goths?

It is one thing to try and understand your own mind, but what about when you try to understand other people's – in particular your own children's minds? You know you have already screwed them up with those genes that may have passed on some of the more lumpen parts of your personality, but is there still time to get in their minds and make some reparations?

Sarah-Jayne Blakemore is a neuroscientist who has specialized in the teenage brain. Maybe you have a teenager in your house, or maybe you *are* a teenager. The teenager versus anyone who is nearby can lead to explosive conflict. The teenage years and the behaviour witnessed are often described as 'just a phase', but it is not merely a colloquial phrase; it is a very real phase of neural development. By understanding what is going on scientifically, we may be able to appreciate that what can often be seen as moodiness and pointless conflict is a very real part of nature, and is necessary for development.

Now that Sarah-Jayne has teenage children, she has found that her area of speciality has very real-world uses for her. Knowing about teenage development, and particularly teenage brain development, helps when it comes to being a bit more sympathetic, understanding and supportive of difficult behaviour in the teenage years. She has found it useful to know that there are adaptive biological reasons why teenagers behave the way they sometimes do.

When she was an undergraduate in the 1990s, Sarah-Jayne was taught that basically nothing changes during the teenage years and that the brain is fully developed by late childhood.

When she first applied for a grant to work on teenage development and the brain, her application was rejected. All the reviewers said, 'Why on earth would you want to study teenage development?' At that time if you wanted to study development, you had to study babies, infants or children. Over the past twenty years, research has shown unequivocally that the brain and many different cognitive processes and social cognitive processes undergo substantial and protracted development throughout childhood, throughout adolescence and right into early adulthood.

Sarah-Jayne believes that this gives us a different perspective on what it means to be human. 'The fact is that we undergo this hugely long period of development and developing our sense of self, that our sense of self is not constructed early, but goes on being constructed for decades,' she explains. Neuroplasticity allows us to continue to change, but knowing that the brain and cognition undergo this development right throughout adolescence and early adulthood also makes Sarah-Jayne think about how we construct society. We don't really allow second chances; we have a one-size-fits-all education system and, if you don't match up to it, it can seem to be too late.

The need for meaning also really seems to kick in during the teenage years; the music we listen to, the films we watch and the books we read appear to increase their importance, in terms of what they are telling us and what they are saying about the world. I expressed this with a ludicrously solid quiff and an unread Albert Camus book in my outsize-overcoat pocket. I have read the Camus book now and I have said goodbye to the quiff, mainly because I have said goodbye to my hair (thanks again for the genes, Dad – as if inheriting all that moody personality wasn't bad enough!).

The whole point of adolescence is to develop a much more profound sense of self-identity and, particularly, one's social self. Sarah-Jayne says that's not to say that children don't have a sense of self – they do – but at puberty you place far more importance on the features and the characteristics that define you, your music taste or your fashion taste, or which peer group you are hanging out with. This period of life is all about developing your self-awareness and a sense of self-identity.

It's all just your brain's best guess

Professor Sophie Scott, head of neuroscience at UCL, finds consolation in the human condition in the knowledge that, despite the fact that we live in a huge universe that is oblivious to our existence, we keep pootling on anyway. She sees quite a lot of what we do as pretty simple, social-primate activity. We can be resistant to this idea, because we like to be in charge and making our own decisions. If we form a dislike of someone, we don't like to think it is perfunctory, without deep reason, and as basic as the way a chimpanzee might react. 'The older I get, the more I look at human behaviours and just see social primates in action, being engaged by the joyful and being repulsed by terrible things in exactly the same way,' Sophie reckons.

Pretty much any time we find something that humans can do really amazing, after a little well-aimed research you'll find an animal that can do it in one way or another. The only thing that Sophie thinks is still standing for human behaviour alone is our ability to contemplate multiple meanings. For us, an object can be a hand-axe, a piece of art, a paperweight, and so on, depending on context, time, place and our thinking about it. There is a depth of potential meaningfulness, purpose and interpretation – and that, Sophie thinks, is the last bit of exceptionalism left standing for humans.

Science is only one of the ways that we question what it is to be human. Philosophy, literature and art all interact with the question of what we are. Whilst people are less likely to be offended by art dehumanizing us than by science, they have found plenty of other ways to be offended by art – in particular contemporary art, which is burdened with reactions of multiple varieties of 'Sheesh! My five-year-old could do that.' That is something science doesn't have to deal with so much. 'Call that an effective particle collider? My five-year-old could make one of those out of egg boxes and an old pair of tights!'

In another way, though, Sophie sees us asking absurdly basic questions of the world around us, and then trying to map really complex things onto the world. She believes it is important to be humbled by the much bigger picture. It is only relatively recently, for instance, that we have had models of perception showing that we do not all experience the world in the same way. We believe there is one objective truth about it, and we believe we are the ones who have seen that truth, and yet a lot of what we are doing with information from the world around us is making our best guess, though that guessing game is often concealed from us.

Nature is detecting nature. Our perception may seem direct and unequivocal, but it is not. It is interpreted through expectations. When there is a mismatch between expectations and what we are shown, that is where things get interesting. Sophie explains that we are not perceiving it directly, yet we are also totally embedded in it, because we are embedded in the biology of it. Your vision is working on the basis of chemistry. Hearing is working with basic physical movement. Smell is bits of a thing that you are scenting hitting the nasal mucosa. It is both extremely abstract and completely embodied physically.

Sophie says there are incredibly strong cultural differences that influence what has been thought of as the very basic units of how the brain works and how perception works within that. The Himba people of northern Namibia, for example, are not susceptible to some of the really simple visual illusions that trick many people in the West. One visual trick involves circles of the same size being surrounded by other circles, which confuses our judgement of their actual size. You know the two circles are the same size, but circles on the edges makes them look different. The Himba are not troubled by this at all. There is no optical illusion. Their ability to pay attention to a target and not be influenced by distractors around that target – one of the most basic models of the way attention works in humans – means that the illusion doesn't work on them. They are much better at paying attention to things, and ignoring distractors, than Westerners. 'So that kind of thing, when you think about it, suggests that at one level human brains do not work out of the box,' Sophie explains. 'It doesn't mean it's all up for grabs, but if you'd asked me ten years ago, I would have said that stuff about attentional processing – that's just how human attention works. You would find that anywhere in the world.'

One of the most useful things Sophie has learnt for our day-to-day existence is how exceptionally plastic and flexible human memories are. Memories feel incredibly real, but they aren't, because they are affected by many things that go on around you. Even more disconcertingly, the more you revisit a memory, the more it changes, though its veracity for you remains. I find this highly disorientating. Each time I go back, am I making a memory worse or better? I have an annoyingly good shame-memory. I am plagued by memories that make me blush and recoil. Are they becoming more repugnant with each revisit? Am I making myself more of a dick each time I experience that memory again?

One of the great TV quiz scandals occurred on *Who Wants to Be a Millionaire?* when a contestant was apparently helped by some assistive coughing from a partner in the audience. Many people remember watching that episode. The only problem with that memory is that it was never aired. Sophie was surprised when she saw a TV drama based around the incident to find out how her cast-iron recall of seeing the show was a myth.

It is not pleasant to discover you have been innocently lying to yourself. Many people in Britain who lived through the Second World War remember how united they felt when they sat around their wireless radio and listened to the empowering speeches of Winston Churchill, especially his 'We shall fight on the beaches' speech, except that it was never broadcast during the war, it was only heard in the House of Commons. It was reported and later recorded after the war, but never aired during it.[4]

While scientific research may be taking your memories, removing your certainty and promoting doubt, those memories of larking about on sunny beaches, or a first kiss or an avowal of love, are still real, even if you have changed the camera angle, some of the supporting cast and the ice-cream you were eating.

And the next time you are in the full flow of an argument, you can agree to differ a little earlier, as there is a good chance that you are both wrong. There is room for forgiveness on both sides.

Is this my hand I see before me?

Hayley Dewe is a doctoral researcher at the University of Birmingham and she showed me directly that my brain is taking a lot of shortcuts to give me a picture of the world, and that even my sensory truth was not as sensible as it should be. It was in her laboratory that I experienced the rubber-hand illusion. It is a nifty trick whereby, if you place your hand out of view and stare at a rubber hand positioned roughly where it might be, your mind makes the supposition that the rubber hand is your real hand.* Since I took part in that experiment there has been further research, where it appears that people like me, who can be duped by a rubber hand, are more likely to be susceptible to hypnotism. This is why I did not buy a ticket to see the Great Mesmo at Blackpool Pier, because I did not want to find myself eating an onion that I thought was an apple, making love to a chair or assassinating the Ukrainian ambassador while singing the songs of Elvis Presley (all very possible outcomes of hypnosis, I am told).

Hayley says that papers on the rubber hand suggest a high suggestibility and the potential to be easily swayed in those who experience it, which is bad news for Hayley as well as me, as she

---

* Sorry if the details on the rubber-hand illusion are scant, but I wrote more lengthily about it in a previous book, although I am pretty sure it had a different publisher, so I'd better not plug it on these pages.

has also experienced the rubber-hand illusion and she would like to think that, as a psychologist, she is not easily swayed. She thinks it's strange because she obviously understands the process behind the illusion, and the reasoning about why it could trick the brain, but she can't prevent the sensation happening. The inability of knowledge to override sensation is something I find fascinating and infuriating. You would think that, having set up the experiment so many times herself, when she herself was the participant she would immediately see through it all. Some of her *is* seeing through it; it's just that stubborn other part of the brain that isn't listening to her. It is similar to those people who suffer from anxiety and are fully aware that the worry is not founded on logic, but, however much they try to talk to that anxiety, the anxiety refuses to listen, or maybe inhabits a part of the brain that can't understand a word they say anyway.

How the participant approaches the experiment doesn't seem to affect the outcome. Hayley says she has had people coming in feeling very sceptically minded and critical, and then they do get the illusion, which blows them away. Some people who don't get it didn't expect to, and some people have been looking forward to the illusion of connection and are disappointed when it doesn't happen. She says the main problem with the rubber hand that you have to be careful with is the way you measure it. You don't want to feed ideas to people too much. You have to be careful with trigger words, so should avoid asking things like 'Did you feel your hand swelling?'

The main criticisms of the rubber-hand illusion come with questions concerning how and what you can use as a control. If Hayley goes on to threaten the rubber hand with a needle or a knife, you can physically see everybody's reaction and she can tell straight away who actually believes, because their heart rate increases, they start sweating and they scream or jump, even though it really isn't their hand. This further testing of our brain means we should be aware that our grasp of reality is more tenuous than we might hope.

It is the precariousness of our brains, though, and how we experience the world that alarms Hayley. When she was an undergraduate, she was startled in a lecture when she found out that the brain has quite a high capability of going into a seizure at any point. Our brains are on a seizure knife-edge. The idea that a neural storm could fire up at any time did not rest easy with her anxious mind.

Has understanding the human brain increased or decreased her anxiety? Hayley thinks the anxiety might have increased, although that increase is probably more because of age and development and awareness. It is due to the knowledge of what different mental states that she is experiencing actually are. 'This is a depersonalized state. This is an anxious state. This is a sleep-deprived state.'

The ability to accurately observe the state you are in, and to scrutinize the causes you think may have contributed to that state, is something that makes me wonder about the pros and cons of knowledge. After having had therapy, and also after spending time with neuroscientists and reading around the topic, I find that I have a greater ability to dissect my mental scenarios, particularly around anxiety, but I have wondered whether such knowledge and self-observation have ended up extending the time that I remain anxious. I suppose at least it's given me something to do with my anxiety – distracting me away from it by diving straight into it. Or maybe the next time I'm in therapy I'll just ask for MDMA and go out and blow my luminous whistle.

Hayley was a very poor sleeper before she started looking into the effects of sleep. She says she still is a poor sleeper, but she now knows more about the effects, the anxieties tied to sleep and the hallucinations or delusions that are linked to sleep. After all this study, she was driven to change her sleep patterns. She tried to force herself to sleep, downloading sleep apps and trying to stick to regular sleep times. After taking part in a sleep experiment, she was told she was going to bed at the wrong time. She attempted to sleep in what was presumed to be the natural rhythm, but eventually slipped into being herself again. The knowledge has helped her, though, with the idea that everybody is different, so she can be different in her sleep patterns. She has relaxed in this regard into acknowledging that she is who she is. Study has not been a cure, but it has brought some peace of mind.

Techno-rapture nonsense

The realization that we are unreliable narrators of our past – that our memories, however exact they may seem, are likely to be inaccurate and to become increasingly inaccurate, the more we revisit them – is discomfiting. This came into sharp focus through the research of cognitive psychologist Elizabeth Loftus, which was prompted by her investigation into the reliability of allegations surrounding childhood abuse. She found a number of cases where healthcare professionals had encouraged their patients who had no memory of abuse to imagine that they had. In books about childhood abuse, she found therapists writing things like, 'Spend time imagining you were sexually abused, without worrying about the accuracy.'[5]

Loftus wondered what imagining such situations might do to the memory. Her research led to the concept of 'imagination inflation'. This she describes as 'the phenomenon that imagining an event increases the subjective confidence that the event actually happened'.[6] What research has shown is that, after imagining an event, this will have little effect on the accuracy of memory if you are asked about it a day later, but if the gap between imagining events and testing increases, then the convolution between memory and imagined memory increases. The ability for professionals with an agenda to play with beliefs that we have about our past is deeply troubling. Loftus has performed research where she has been able to implant very specific false memories, such as kissing a frog or witnessing demonic possession.[7] Our memories are more fragile than we might imagine.

Meanwhile, the experiments of neuroscientist Benjamin Libet suggested that our brain has often made up its mind on our decision-making *before* we self-consciously know what we are going to do. In his experiments, volunteers performed simple tasks, such as flexing a wrist or pressing a button, and it appeared that the brain made the decision before the participants made their conscious decision. This was a worrying result for any of us who might prefer to believe that we are not really post-hoc rationalists of actions that would occur without our conscious participation – observers under the delusion of being in charge, while ultimately powerless (a bit like many democracies). Fortunately, the presumptions from this experiment are still being debated, so keep acting as if you are still in charge anyway.

But these seismic shifts in our potential for being responsible and reliable in our decisions and actions are as nothing compared to the next idea, which is, thankfully, untestable and unprovable – or at least seems to be in our world, whatever world that is.

Consider being told that the entire world around us (and indeed we ourselves) is nothing more than a simulation, which possibly came out of another simulation that came out of another... This is 'simulation theory', and whatever physical reality existed to create the first simulation is far, far away now. It's at this point that Keanu Reeves exclaims, 'Whoa!'

Simulation theory doesn't make computational and cognitive neuroscientist Anil Seth happy; indeed, he thinks it is an ugly idea. Anil studies the brain basis of consciousness. The first time we met, we were recording a discussion about simulation theory with a philosopher who had eaten the contents of the backstage sandwiches with a teaspoon. There was an underlying tetchiness about the disagreement.

Anil sees science as having an aesthetic quality, like art, in that there are ideas and theories that are pleasing to us in their hope and form, but simulation theory is not one of those. How do our brains and minds change if we are living in an advanced digital construct, possibly a universe fabricated by an artificial intelligence? Anil finds the idea unappealing because it doesn't change anything about how we and the world work, and instead is made via a whole raft of assumptions whose plausibility or implausibility you can't ascertain. 'It is predicated on all sorts of back-of-the-envelope reasoning, such as what are the chances we'll get wiped out at different stages? And if we don't get wiped out, then we have to sort of say, "Well, we're more likely to be in a simulation than not.'" Underwriting all this is the assumption that conscious experiences are things that can be simulated and, without that, Anil sees it as a meaningless argument.

It captures many people's imaginations, perhaps because it has a strangeness that fits well in a science-fiction culture filled with superhero movies. It is this idea of the inevitability of creating consciousness, once we reach a certain level of technical know-how, that annoys Anil, because much of his neuroscientific life has been spent contemplating consciousness. He sees those who insist on the likelihood of simulated universes perhaps having too much confidence in the inevitability that if you can program a computer to play Go, one day it will create a cosmos. This confusion is partly down to the ingrained notion that the brain is an information-processing system. Anil defines himself as agnostic, when it comes to the idea that consciousness could be simulated. In the story of consciousness, simulation theory is a plot for a short story or the impetus for a paranoid outburst, but not a useful theory for understanding ourselves.

What Anil has found interesting in his study of consciousness is a growing sense of commonality and a push-back from anthropocentrism. In the ancient-Greek great chain of being, the gods are above us and the animals below; we seek ways to bolster our sense of superiority over that which mewls, yaps or flutters about us, but our conscious experience may not be so separate that we can draw a neat and bold line between man and dogs, cats or pangolins.

Anil has increasingly recognized how important the body is, and how important our nature as living organisms is, in understanding many things about what we are and the way we perceive being an individual self and being alive. He sees the basic aspects of selfhood as drilling down into what brains are ultimately for, which is keeping the body alive. He views this insight as a way of thinking about who and what we are, which should bring us closer to nature and make us feel more part of the natural world around us. The burden of our hankering for exceptionalism can prove destructive.

This sense of commonality – a sense of a conscious experience that, although different in different species, is a shared trait – wasn't something Anil set out to discover, but he feels that it is something he tries to project back into his own life and finds psychologically helpful: to not be apart from, but a part of, nature. It is a good trade – to give up a little bit of narcissism and place ourselves more deeply in the fabric of the universe and of nature.

He sees the paranoia of artificial intelligence enslaving us all as coming, to some extent, from simulation theory, and from the assumption that consciousness is associated with intelligence; that when something crosses the threshold of intelligence, it becomes conscious. Anil views this as a symptom of a pernicious human self-centredness. Being intelligent gives us more ways of having experiences, but if consciousness is more to do with being alive, it is correspondingly less to do with being smart. This enables us to see ourselves as existing along a continuum, fitting into a bigger arc of how science has led to our reflections on where we are in the universe.

We have always gone from the perspective of seeing ourselves at the centre, to a deeper understanding. Copernicus showed us that we are not at the centre of the universe. Darwin revealed that we are not fundamentally different from all other animals, but have a common ancestor stretching back into deep time. This should be an enriching experience, as it extends our family. The third stage of this is that conscious experiences don't make us special or distinctive, either. They are also part of the natural order.

It doesn't mean that we are not special; there are still many things that make humans very different, but we are not special to the point of a unique divide between us and the rest. Anil's definition of conscious experience is 'any kind of subjective experience whatsoever. For a conscious system or organism, there is something it is like to *be* that system.' There is something it is like to be you or me, but not anything it is like to be a shoe or a pebble. He gives room for there to be something it's like to be an insect or an amoeba, but probably not a bacteria or a virus. He has a scale of consciousness.

A really simple animal might experience happiness or sadness: things are good or things are bad. A slightly more complicated animal might be able to experience disappointment: 'I thought things were gonna be good, but they're bad.' Up one more level and an animal could experience regret, which requires more complex cognitive processes, a need to entertain counterfactuals. And then there is something that Anil suffers from – anticipatory regret. 'You know if you do X, it's gonna turn out bad and you're going to regret it, but you do it anyway.'

For Anil, one of the beauties of science is that you can put things into a larger context, and through that you can see that maybe you are less personally responsible for them. 'I think this is one of the one of the more existential things that science can offer. A lot of unhappiness comes from over-emphasizing our personal role in the history of our life and in the universe,' he continues. 'Some people might find the vastness of the universe – of space and time – to be quite threatening. Others, including me, find it extremely liberating that this is so much grander than we could possibly imagine. I don't have to feel bad about my own little parochial corner of whatever the universe really is.'

In my conversations with neuroscientists I have become increasingly aware of my cognitive biases and my unconscious biases, and that our biology may make us more predictable to such bias. I may not be as free an agent as I once thought I was. Whether free will is an illusion or not, perhaps it is best to find a balance in your head: enough to feel responsibility, but not so much that you have to shoulder the burden of all that you think, do and say. This might have a positive influence on our sometimes crippling feelings of regret. Regret requires us to believe there were alternatives to our actions, so as free will thins, our responsibility decreases and therefore our regret should become a lighter burden. Perhaps Sartre was wrong: we are not condemned to be free, we are sentenced to believe we are free.

Then perhaps we can grasp the recent understanding of the brain and its ability to change – and to change at any age. A few decades ago students were taught that, after the age of sixteen, you lost brain cells. Your decline began just as you started trying to take on the world. Now we know that not only are new cells being made all the time, but the brain is remarkably plastic even without having new neurons, so that when you learn a new skill, the brain is actually being changed. Our potential has increased with this knowledge. Okay, there might still be a little more debate about exactly how much *you-ness* there really is, but your brain is creating one hell of a cover story, so enjoy it when you can and blame your unconscious, subconscious or amygdala when you can't.

I was taught that the human brain was the crowning glory of evolution so far, but I think it's a very poor scheme for survival.

KURT VONNEGUT

CHAPTER 9

Reality, What a Concept[1] – Can Anything Be What It Seems?

Reality is nothing but a collective hunch.

Lily Tomlin

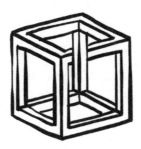

The discoveries of quantum physics and our developing neuroscientific and psychological understanding of how we perceive the world have made our conception of reality a fragile thing. Seeing is not believing, but neither is hearing, tasting or touching. The more science books you read, the more you see the phrase 'the hidden underlying reality of the world'. It's enough to make you paranoid, especially if you have just confronted the possibility that you are a post-hoc rationalist. Reading too much Philip K. Dick doesn't help, either. We have become aware that not only do we all see the world differently, but that none of us are really seeing the world at all. Instead we are receiving an image of the world, and our mind is filling in the gaps with guesses and presumptions.

As our understanding of our understanding of the world has developed, artists have created work that is less and less definite – pictures and plays that don't offer a single meaning. There is no definite answer to 'What's that all about, then?' This is what infuriates those critics of contemporary art whom I mentioned a few pages back. They want a tree to look like a tree, a horse to look like a horse, and a breast to look like a breast. They want to know who the good guys are and who the bad guys are, and they want to know how they are meant to feel at the end of it all. The older I get, the more I like what I look at to be uncertain. Watching David Lynch's *Mulholland Drive* in a supremely jet-lagged state in a San Francisco cinema was fine for me, even if the next day I couldn't be sure what I had dreamt and what I had seen. Why should art offer something that is easy to grasp, when life so often doesn't?

The older I get, the more I enjoy Samuel Beckett. I look around an auditorium and usually presume that I am not understanding it with the depth that the others are, but even so I like to sit back and see what happens. One of the joys of watching *Waiting for Godot* being performed is waiting for the point where the audience realize they can laugh. The play is seen as one of the great pieces of art of the twentieth century, so it can be watched with a worried reverence by people terrified that they may not be understanding whatever the thing is they are meant to be understanding. In my experience, there is usually a breaking point where the less certain are persuaded that it is okay to laugh.

Vladimir: What do we do now?

Estragon: Wait.

Vladimir: Yes, but while waiting.

Estragon: What about hanging ourselves?

Vladimir: Hmm. It'd give us an erection.

Estragon: (highly excited) An erection!

Vladimir: With all that follows.
Where it falls mandrakes grow.
That's why they shriek when you pull them up.
Did you not know that?

Estragon: Let's hang ourselves immediately!

Waiting for Godot has been described as being a Rorschach test. There is not one meaning in it delivered to you by the author, but instead you play an active role in deciding what it means for you. This is exactly the sort of work that delights any academics leading courses in literary theory, as you can simply keep interpreting what you have seen over and over again. The observer is an active participant. There is not one Godot. As the previous chapter suggested, reality is slippery and probably not ultimately available to us, although we can find the best workable model. Some things need to be seen to be believed; others need to be believed to be seen. As we see from our experience of time and experiments in visual perception, reality is more flexible than our common sense may wish it to be.

We may accuse people of 'not living in the real world', but who does? We can only hope that we have enough agreement on the model of the world to deal with important issues pragmatically.

Let's hope that when I see a grizzly bear, you see a grizzly bear too, or our camping holiday in the Rockies may be brief.

Personal reality is hard to measure objectively. The subjectivity of pain is an example of a reality that cannot be shared. However much we empathize, we can only guess at the pain someone else is feeling. Pain specialist Lorimer Moseley has said that the first thing to take into account is that 'pain is always real, whatever causes it'. If someone tells you they are experiencing pain, you cannot dismiss it with a simple, 'Don't be so silly, of course you are not in pain. Your screaming agony is an illusion. Now pull your arm out of that grizzly bear's jaws and let's get this tent up. Oh, you're saying it's not a grizzly bear, it's a wood-chipper. I don't think we'll ever see eye-to-eye on this holiday.' In his lectures, Moseley uses an example of pain expectation from a *British Medical Journal* report. A construction worker came to hospital in agony. A large nail had gone straight through the middle of his boot. His pain was so excruciating that maximum doses of morphine were given. After an X-ray, an interesting discovery was made: the nail had not pierced his foot at all. It was lodged in the gap between two toes. His belief in the brutality of the accident had created a very real sensation of the agony that should have occurred from such an incident.

Sometimes our misperception may solely be down to a mistaken belief in a situation, and sometimes it can come from damage to the brain. One of the most disturbing situations of alternative reality is Capgras syndrome, or delusional misidentification. This is the denial of the reality of a close family member or loved one. It is similar to the situation imagined in the B-movie classic *The Invasion of the Body Snatchers*, but in this case your life partner has not been swapped for an identical-looking pod person (within reasonable boundaries of certainty). It occurs when there has been damage to the area of the brain that connects visual input and emotional response. When you see your loved one, the lack of emotional response means that you do not believe it can really be them. If you speak to your mother on the phone, you do not believe she is an impostor because the auditory emotional connection is intact; however, should she walk into your house ten minutes later, you cannot believe it is really her. Like so much in the way we interact with the world, the emotional reaction that accompanies seeing those we love is unconscious. Its existence becomes apparent only when it is lost. Most disagreements about reality are not this devastating, but as we understand more deeply the foibles and fripperies of how we experience the world, it is another example of how it loosens the grip on our certainties.

In the book *Stranger Than We Can Imagine: Making Sense of the Twentieth Century*, John Higgs looked at how early-twentieth-century scientific revelations about the relativistic nature of time, the quantum behaviour of particles, affected art and society. He described it as the total loss of the omphalos, or centre. Not only could the centre not hold, but these revelations showed that there was no centre in the first place. In science, the concrete objectivity of a Newtonian universe was sacrificed for a new subjectivity. The solidity of the clockwork universe was replaced by something far more ghostly. Continental philosophy further contributed another form of cultural relativity. All forms of truth were destabilized.

Rather than leading to a culture full of healthy doubt, the multiple ways in which reality may be questioned has led to a proliferation of dogmatic ways of belief in preposterous truths. Rather than making reality questionable, it has made it dispensable. Rather than a civilization that questions its reality, we have an increasing number of people who cherry-pick reality, accepting only a vision of reality that is how they wish it to be, and no other. Perhaps the most bizarre new reality is the return of the flat-Earth movement. This movement is not merely a gathering of loveable eccentrics, but can also be angry and pugnacious.

The documentary *Behind the Curve* chronicles some of the proponents of the new flat-Earth movement. They do not appear to be stupid, and they are often ingenious and inventive, yet they believe something utterly ludicrous. If this was merely down to stupidity, it would much easier to understand. In the film, one proponent of pancake Earth decides to test his theory with a gyroscope. He correctly surmises that if the Earth rotates 360 degrees in twenty-four hours, then a gyroscope will shift its position by fifteen degrees every hour. For some reason, his result is that the gyroscope moves fifteen degrees every hour, which can't be right, as the Earth is flat. So he decides it must be something to do with the sky and therefore reworks the experiment. Same result. He will not rest until he has found a way of his experiment proving what he wishes to believe. This is the opposite of what science should intend to do (although the history of science is not without people considered to be legitimate scientists who have mangled experiments to get the results they wish for).

One of the problems of reality is just how many realities there seem to be at any given moment. You are probably in at least two, right now. There is your reality as you perceive it, and there is the scientifically testable reality, which can be turned into equations and whittled down to a more objective reality. Some people might wish to dismiss the former as a human frippery, but our fripperies are persistent and dominant, and this frippery is the lens through which we see it all. However immeasurable the reality we perceive may be, it is the one that dominates our day, our moods and our judgement. If we needed to test the veracity of our perception from the moment we woke, we would not get much further than our pillow, but the measurements from eye to bedroom ceiling would be increasingly accurate.

How strange to have a reality that you are in and a reality that you experience. We are immersed in an invisible reality. With scientific knowledge, maybe we should say there are three realities: my internal reality, the consensual reality of the outside world and the underlying world of quantum behaviour – of atoms that are mainly empty space, with the potential of it all being a hologram anyway. The suspicion that everything is not as it seems is no longer necessarily a paranoid position. Perhaps the immediate leap to 'and thus the trans-dimensional overlords are controlling the world, and all diseases are made in a laboratory on Saturn by goat-legged lizards' is, therefore, comforting to some.

When sleep-deprived and not using my insomnia to travel to alien planets, when my muscles seem neither asleep nor awake, but hazy, I drift off into my atoms. I become detached from the scale of the world as I usually experience it. I think about all the particles that make me. How individually, each atom – some carbon, some hydrogen, some oxygen and the rest – is thoughtless, but for a short time they are part of a feeling-and-thinking collaboration. I look at the objects around me again, at this point a grandfather clock, an empty picture frame and a bottle of Welsh whisky, and I am aware the atoms in those objects may have spent, or will go on to spend, time as part of a system that thinks and feels – especially the Welsh whisky. At this point I feel like atoms going through a brief phase of delusions of grandeur, before my less animate and less conscious destinations. As Jan Westerhoff writes in *Twelve Examples of an Illusion*, 'An illusion is not something that doesn't exist, but something that is not what it seems.'

In Tibetan Buddhism there is something called a Tulpa, which usually takes human form and is created by someone's imagination. Am I my own Tulpa? Westerhoff writes about a Tulpa as being 'different from other magical beings in that it can be seen by other people as well, and in that it may also acquire a certain degree of independence: When the creator wants to dissolve the Tulpa, it may not disappear immediately.'

Just as quickly as I get lost in my mystical whimsy, I return and find that I seem to feel everything for a moment: the hair growing out of my upper lip; the faint ache in my shoulder from yesterday's lockdown session with the kettle-bell to combat my increased cake consumption in isolation; the socks against my toes; the faint pressure in my bowels that I don't need to pay attention to yet. I am back... I think. I know that my other reality was my imagination and sprang from restlessly reading about atoms and then reading some Herman Hesse.

Is reality psychosomatic?

'When I saw *The Matrix* at a local theatre in Slovenia, I had the unique opportunity of sitting closely to the ideal spectator of the film, namely, to an idiot,' wrote the philosopher Slavoj Žižek.

The Matrix did for Plato's Cave what *Die Hard* did for sales of grubby vests. It was a crash-course comic-book guide to philosophical arguments about the nature of reality, but with guns... lots of guns. On one level, the film is an adaptation of Plato's Cave – the idea that all we ever experience is an illusion – and on another it is a Hong Kong action remake of the New Testament, with Keanu Reeves playing the part of karate Jesus.

The Matrix offered a simple solution to the conundrum of Plato's Cave, in which the inhabitants believe they are experiencing reality, when really they are only looking at a shadow of reality on the wall in front of them. Chained to the ground, they are unable to turn round and confront the truth. In *The Matrix* you took a pill and experienced reality, which, it turned out, was far worse than the fictional existence that the aliens provided for you as they sucked the energy out of your skull.* Unfortunately, seeing the levels of reality of the structure and content of the universe that are not immediately apparent to us is not available in pill form. It is available in equation form and in huge wipe-boards of numbers and symbols, all of which is far harder to digest than a single pill. Big pharma has not made available reality by prescription. It is not the aliens or sentient

* This is all based on the first *Matrix* film; I don't know what the second and third ones told you, but I didn't like the look of them. It appeared to show that reality was a rave, and that is not my cup of tea *at all*. I have never owned a luminous whistle.

algorithms that are hiding reality from us, but the universal laws of physics.

According to classicist Natalie Haynes, Plato's Cave may have had a prescription cure – although via the optician rather than the pharmacy. Natalie doesn't like Plato's idea that we live in a failed copy of an ideal. She doesn't feel disappointment when she sees the real world, in the way that Plato must have. She is attracted to the classicist Edith Hall's theory that Plato was very short-sighted, and that is why he found the real world so disappointing. He literally couldn't see it very well. Plato found the life of the mind more satisfying because he could see things clearly in there. So, by default, I wonder whether Aristotle's optimism may have been a product of his superior eyesight.

Science is the search for an objective reality by a creature with a subjective view of the world. The objective world beyond us means that we cannot believe or perceive ourselves out of certain situations, however much we feel that the laws of physics are open to interpretation. We can debate reality, but some of its actions are inescapable.

Geneticist Paul Nurse has a deep understanding of the quest for scientific knowledge, but he wonders if we may have reached the limits, in terms of our ability to comprehend. Are there limits to how far we can dig down into reality? He believes that in a very fundamental, philosophical way, which is beyond his comprehension, the human mind has taken us into a place where it isn't natural for us to go. When you are hunting antelopes on the savannah, you have a sense of time and three dimensions of space, and you throw your spear and it all works well. Is there a depth of reality that we can never see and will never know because it is invisible to us? As we aim our spear, we don't have to think about relativity. The same curiosity that generated the ability to invent a spear, plan a hunt and kill an antelope, when taken further and further, has eventually led to relativity and quantum mechanics. It goes beyond the scale and experience of the world we view. We can write equations that don't quite make sense in our real world, yet have extraordinary predictive power. Their reliability can be tasted, but at a level that is quite beyond the evolved parameters of experience.

'We show that the world is not straightforward, but we're looking at it and we do our best,' Paul explains. 'Physics is the most wonderful discipline, but it isn't in our everyday world. And we haven't cracked it yet, because we can't obviously relate gravity to the forces operating at the level of the atom and below. And I think biology will end up there too.' For Paul, it is humbling. 'Somehow, we've got this wet grey stuff doing chemistry between our ears. And somehow it produces this – this thinking. I mean, this is a wonder of the world. We wonder how we can produce something that is self-aware and can love, then can feel pain and sadness and all these things. But it also is wet grey stuff doing chemistry and thinking about theories of relativity.'

This seems to be a universe of different realities, and we have evolved to survive by interpreting the world at one level. Dealing with the possible reality of being two-dimensional projections from somewhere else in the universe, as black hole research suggests, is not a reality that we need to worry about. It changes our understanding, but it doesn't alter our weekly shop.

Similarly, the understanding that atoms are predominantly empty space doesn't change the need to get on and repair that bookshelf above your bed. Should it fall, the atoms will not give your head a free pass. We are Newtonian creatures on a Newtonian planet, from a neurophysiological point of view, and even if physics now shows us that there is an underlying reality beneath the surface, being aware of individual electrons holds no day-to-day advantage for us. Our senses have evolved to deal with the most relevant issues. Concentrate on the empty spaceness of your atoms as much as you want; you will still find it far easier to get into the next room by opening the door rather than trying to walk through it.

Oh, so many realities!

With all these realities, it makes me wonder if there really is an ultimate reality. Physicist Brian Greene thinks so. He explains that reality is a model made from the laws of the universe and the beings that measure it with their nervous systems and instruments. It is with those ideas of reality at a quantum level that the quandaries really kick in, for many of us. We have been told that light behaves as a wave and as a particle, and that a single electron interacts with itself and appears to be in at least two places at once, unless we are looking at it, and then it behaves as one particle. It is as if the universe is made up of tiny disobedient children who create havoc every time you turn your back, but are all sitting primly and obediently at their desks the moment you look at them. It feels as if the universe is in utter anarchy every time I blink, and then regains its composure.

Next time you lose your keys or find a proliferation of single socks, perhaps the problem came about when you stopped observing them for a moment, and then the subatomic particles got mislaid during their unobserved hullabaloo. Basically, there is one law for us and a totally different law for the subatomic particles that make up us and the rest of the universe.

Brian explains, though, that the quantum universe is not a reality, but a series of probable realities that only become reality when captured by measurement. The possible realities that we talk about, when there are observers, remain multiple realities if observers do not exist. Or is there a single deep reality, possibly elusive to us with the biology that we have, but in existence somewhere? For Brian, it depends on your definition of what constitutes a reality. As a physicist who is deeply immersed in the quantum paradigm, he would say that reality is the probabilistic description of the positions and the motions and, more generally, the quantum state of the particles. This is a reality to him; when anything measures reality, it affects reality.

Brian says, 'It can snap out of this probabilistic quantum haze and assume a single definite configuration or state of being, but that, to me, is just an interesting event that can take place in a quantum universe, but the quantum description of reality in terms of a range of probabilities dictated by the equations of quantum mechanics. That is reality. That is the rock-bottom reality.'

I have already mentioned black holes and their size and power, and that they might stretch you into a length of pasta, but with this enormity there are also revelations about the very small, and that brings into question some dearly held beliefs about our reality. An idea that springs out of black holes is complementarity. This suggests that you can't always know everything simultaneously, but you can know one description and then you can know another description, and they are complementary. Falling into a really huge black hole, your pasta-stretched demise may well not be instantaneous. You might be swimming about for a year. From outside the black hole, you are smeared on the horizon. As long as there is nobody who can simultaneously observe both things at once, everything is okay, but this idea becomes problematic, if it isn't already. Stay with me...

There are two different versions of reality occurring. The idea is that both things can be real, as long as there is no contradiction. The problem that occurs – an issue that would never crop up when hunting woolly mammoths – is that this might lead to us giving up on the idea that things, including us, are located in space. This is where the reality gets weird again, or at least weird from the perspective of us, but possibly humdrum in terms of inevitability in a universe such as ours.

Theoretical physicist Sean Carroll says that the idea of a location in space is just a good approximation – that it suits our circumstances. When things get extreme, it moves to being somewhat true, but not the fundamental reality. Fundamental reality is something non-local. You do not have a location in space, you are spread out in some quantum-mechanical wave function. This is where the idea of holograms comes in, as this is where we lose our roundedness and become two-dimensional. Is this three-dimensional world you see ahead of you: that mug with the cold tea, that bust of Leonard Nimoy on your desk, that desk – all simply two-dimensional, a projection from somewhere else?

If it is not bad enough that Copernicus un-centred the Earth, it now seems there is a possibility that the whole of space and time, as we think of them, is gone. You believe you know where you are in space and time, and this reality might fit your needs, but it might not be reality. This goes beyond even the spooky idea of quantum entanglement. Should you have got used to two particles communicating instantaneously, however far apart they are, it gets even trickier when gravity gets involved.

Quantum mechanics says that your observations reveal a tiny smidgen of reality, a keyhole glance of truth, but not underlying reality. According to Sean Carroll, underlying reality doesn't put space and time in a central location. It emerges out of something deeper that we are still trying to figure out. With the twin realities of falling into the black hole, the observer's experience and the traveller's experience, on a particle level, this is not a problem. Particles are comfortable with being in two places at once – it is what they are used to. Our conscious experience is not in two places at once; just the particles that we seem to be.

This reminds me of when the theoretical cosmologist Janna Levin told me, 'Every conversation should have something nobody understands, out of respect for physics.' When I hear a physicist explain the possible contradiction of being smeared on the surface of the black hole but also being in the black hole – two truths, depending on the position of the observer – my mind thinks it understands, but now that I have typed it out and you have read it, my forehead is throbbing, trying to comprehend it all. It is at times like these that I feel I should simply stick to one reality and forget about the other two, at least for the time being. Then the throbbing of my head is forgotten, and I throw myself right back into it and try to understand the quandary all over again.

For some, there is comfort in knowing that it appears to be impossible to have a direct contact with reality. On a second-by-second basis, our brain has evolved to edit our reality to a manageable chunk. If I look at a painting – whether it is one that is meant to be a tree, or a confabulation of a thousand fish being propelled out of a Venusian volcano into the mouth of General de Gaulle – I only see what is centre-frame: what I am focusing on. The rest of the picture has an illusion of being a complete image, but my mind is constantly filling in the gaps where my sharpest senses cannot go.

Similarly, with sound, if I focus on the birdsong that I can hear, I can't really be sure what the radio presenter is saying about the listeners who have sent in stories about their most embarrassing tattoos. If all of reality flooded in at the same time, I would be a gibbering wreck. The whole picture is an impossibility, but my brain is still on high alert, should one of my senses pick up information that might be useful to my survival.

In a busy room of conversations, you may be focused on your friend's anecdote about the bass-fishing trip in Denver when his friend was killed by a bear that he thought was a squirrel, and not consciously be hearing anyone else's conversation, but should someone say your name, your senses will be alerted to that and your auditory senses will immediately stop concentrating on Large-mouth Bass and focus on whatever is possibly being said about you. Should it be a false alarm, you will soon be back to listening to the bass anecdotes, though perhaps your brain will then distract you as you work through a series of alibis to offer you an excuse to move away.

Don't worry – it's all in your head

All the reality we experience is inside our head, but it creates a layered sense of the exterior. I reach out and touch the sunflower, and my interior world gives me a sense of the outside. All my sensations are actually within my skull. This why you can create all sorts of illusions if you directly stimulate the brain or put your thumbs into your eyes (not too forcefully) and place pressure on the optic nerve – here is your starry, starry night. Stimulate the brain in the right place and you can smell bread or roses, when none are available to us in 'reality'. The more we find out about how to create reality through direct stimulation of the brain, the more the brain-in-a-vat theory seems feasible. How complex would the electrodes and physical prods directly to the brain need to be to create a whole world while we – eyeless, earless, skull-less and body-less – float in organic gloop?

The question is not so much whether it is possible for us to be brains in vats, but more: why would this Cartesian evil mastermind want to create this farm of brains? Why would he be demonically chuckling away as he prodded our brain so that we experienced a 'reality' of being bored on platform three of Slough station, or of removing the next door neighbour's cat excrement from our rhubarb patch?* Though I suppose, on both occasions, these are tiny hells, so maybe I am arguing against myself.

The musician Brian Eno is always creating new worlds and working out ways of changing our experience of music and physical art forms. He believes that you can't disqualify an experience simply by saying, 'It's all in your mind', because of

* Overly specific, but relevant to this particular Tuesday.

course all our experience is all in our mind. There isn't that much that is objective. One of his essays began, 'How do we change our minds?' He considers that our culture is now a post-enlightenment culture. We tend to think that minds are made up on the basis of evidence and logic, and rational discussion and examination. In some situations that is how we change our minds and how we make decisions. More often than not, though, that is not what happens. We do it on the basis of feelings and hunches, and some idea of what the consensus around us is.

What we believe about the world can come from very vague processes, but we can take very definite actions based on this. 'Rather than saying, therefore, those are not good decisions, we have to say reason and all of the wonderful products of logic and deduction are only available in certain situations and can only work in certain situations. The rest of the time, it's much messier,' Brian tells me. He sees our notions of reality and truth as relying on a kind of mixed stew of feelings, guesses and a sense of looking over our shoulder to see what other people are doing, and he thinks often that is not so bad. The view of hardcore science may be that this isn't a proper way of making a decision, but Brian remembers a scientist saying to him, 'Philosophy is just science that hasn't been done yet.'

Is Pan in the woods?

If I can be befuddled with the reality of the laws of physics, maybe I can do better with imaginal reality. I was introduced to the idea by the neuropsychologist Paul Broks, whose book

Into the Silent Land beautifully and sometimes tragically elucidates the problems of reality for people whose brains have become damaged through accident or illness. Imaginal reality is a convergence. We have a notion of reality and we have a notion of imagination, but they may not be as separate as we like to think. They may overlap in lots of interesting ways, and a stance that relies purely on science might block off a lot of potential ideas and experiences. Paul is more fascinated by why people have the reality they experience than by the deep reality that might lie beneath. He discovered ideas of imaginal reality reading psychologist James Hillman's book *Pan and the Nightmare*.

Paul explains that the human mind has evolved through cultural evolution. The world we are brought up in, and the information fed into us, leads to us having expectations of the world. Because we have expectations of the world, sometimes the picture that we see of it is created from our sensory perceptions, but those perceptions are coloured and shaped by what we believe we should see. How do our landscape and the events in it change, with new information and expectations? How do our beliefs alter our experience of reality?

Take, for example, our beliefs in gods, or God. What is the qualitative difference between the gods of ancient Greece and the God of the twenty-first century? Do we experience things in the same way as people 4,000 years ago? Almost certainly not. It isn't that our brains have evolved and we are wiser now, due to adapted skull hardware, but because our cultures and the knowledge in them have changed. When I walk through the woods, how different my reality must be from those who walked through a woodland believing in imps and witches, where Minotaurs may have dwelled, where a panoply of gods might have been watching, some of which might have taken the form of a fox or a jackdaw. The culture we are born into creates the reality tunnel that we walk through.

Paul's research into imaginal reality led to him connecting those ideas to people who experience sleep paralysis. 'When you talk to sleep-paralysis sufferers, they have these hallucinations that are often fully fledged hallucinations. They experience people as being real people in the room and they feel their physical presence,' he explains. The twin realities that we live in – the experiential and the scientifically testable – sometimes they grow further apart. It is hard to explain someone out of an experience they are convinced of.

Is that how the Greeks experienced their gods? It is quite usual to get ideas and have no idea where they came from. Ideas usually strike, rather than form slowly through concentration. They do not seem attached to you, even though they surely came from you. What if it is more than an idea? If what strikes is a hallucinated presence that seems real, and you link those thoughts to being produced by that hallucination, then you may have something akin to the way the Greeks experienced their gods.

'When someone like Achilles has to make a big decision, he doesn't make a decision; some god appears and tells him what to think. Even Socrates had his demons. If he took the wrong decision, his demon would say, "No, hang on a bit. Now, you don't want to do that." So we've kind of moved through that – we've moved through various stages of cultural evolution,' Paul explains. We have this idea that we are autonomous possessors of an encapsulated mind, and he thinks that maybe it hasn't always been that way. We get glimpses of it, moments of great creative insights when the thinker has felt embarrassed to claim them as their own.

Paul wonders if there is a realm of imagination that goes beyond what we normally think of when we actively imagine things. In some circumstances, those imaginings can even lead us to conjure external beings, as with a sleep paralysis. It is not always enough with reality to say, 'There is a perfectly rational explanation.' Scientific reality cannot fully explain the depth of emotion that you have experienced in a sleeping or waking dreamscape. If you have a really terrible row with your partner in a dream, you cannot always shake it off at the alarm call, saying, 'And it was all a dream'; the suspicion and agitation in your fevered imagination may hang around, unwelcome. In a world that aspires to being reasonable, we are still navigating the way of acknowledging a personal reality and accepting that our mind is the creator of it.

Paul came across the case of a woman who found herself being attacked by a witch. She was in someone else's house and there was a mirror on the wall. Looking in the mirror, she saw no witch. She was seeing the veridical room, but this product of her imagination didn't reflect in the mirror. When she looked away from the mirror, there was the witch. Her imagined reality did not imagine as far as the reflection, yet the witch was still in front of her. It shouldn't always take a ferocious witch to remind us that our reality is precarious.

Bring on the tiny horses...

While discussing the reality in our heads and the reality behind that helps create the internal reality and consensus reality, Paul recommended watching *Devil in the Room*, a short film by Carla MacKinnon about sleep paralysis and the visions that accompany it. 'There is a very thin and porous partition between the stuff of the real world and the stuff of the symbolic, imaginary, fantastical world,' we hear, as we watch a stop-motion puppet lying in a febrile state.

Carla has experienced sleep paralysis and the vivid reality that has accompanied it. She has also been a lucid dreamer – aware and controlling people and events around her in that dreamscape. She has felt the spit on her cheek of crouching night-hags. When your imagined reality is so close to your sensory waking world, how can you delineate the physical world from the mental world?

Carla always had quite strong dreams. There have been periods in her life when she felt as if the reality or the experiences that she was having while asleep seemed as rich and as real as the experiences of the waking daytime. She describes the dreams as epic, often with false awakenings. The false awakening is the creepy conclusion to the horror classic *Dead of Night* and its unsettling story about a ventriloquist doll that takes over the mind of its owner. Sometimes Carla would feel as if she dreamt for ten hours, but the dream was the same scene over and over again – a labyrinthine dream. Sometimes she was aware it was a dream, and sometimes the exit from it seemed quite impossible. The dreams felt completely real. At points in the dream she would realize that she was dreaming, and would start to apologize to everybody in that dream and explain to them that they were not real. These figments of a dream can lead to panicked existential breakdowns. Carla would explain to the figments that they were going to disintegrate and no longer exist when she awoke. She remembers a supermarket checkout woman in her dream who had no particular role, being only a supporting character at best, and Carla had to explain to her that she was not real. The checkout woman started crying and trying to hold on to Carla as she was waking up. Once awake, Carla felt a sense of guilt at extinguishing this existence.

When Carla started making her film, sleep paralysis was a problem in her daily life and she could not find any science that was particularly helpful. The huge gaps in scientific understanding could be conveniently filled by unnameable horrors, which some might describe as an ancient evil conjured up from the pre-language area of the brain – a creature that cannot and will not obey your words. She was living on a spectrum between realities, with one foot in one world of imagination and the other in a physical reality. Although the dreams could be very real, this did not lead to uncertainty when she was back in real reality. There was no way of identifying what kind of difference in quality there was from the inside, so she couldn't have told you in the dream that she wasn't in real reality, because it felt like it, although real reality felt different once she was back in it.

Carla learnt tricks to test reality. She would take a book from the shelf and if she could not read a long passage from it, then it was likely that she was in a dream. She has found light switches are also less obedient to touch in a dream. She spoke to someone whose rational fear, having experienced sleep paralysis, was that this was what would happen when he died – caught between being alive and being dead. Like Carla, in his experience tiny moments of dream consciousness, maybe thirty seconds, could seem to last eight hours. How many seconds might be required to create eternity? Hell didn't seem so far away.

Since making her film, Carla has stopped getting sleep paralysis. When she was aware that she was approaching sleep paralysis, instead of panicking, she saw a research opportunity. She was so hypersensitive to symptoms coming on that it meant she didn't accidentally fall into sleep paralysis without realizing what was happening. This was scary, but she could just sit it out, and eventually the hallucinations stopped. How much had those hallucinations changed her relation to consensual reality? She would sometimes experience small hallucinations during the day, a moment of drifting off, but she would soon be back again. Carla never felt as if reality was at risk. Emotional experiences could continue to hover from out of the dream into her waking day. If there were people in the house with her, she might feel as if she had woken up, and she'd see them standing over her or hear them talking in her room about her. Once fully awake, she would realize those people probably hadn't been standing around or staring down at her with surgical masks on, but there was sometimes a continuing faint suspicion and damage to trust for the rest of the day.

The idea that the most reasonable functions of our brain battle with the agitated parts of it fascinates me. I have seen old people wake up in their bedrooms and find themselves anxious and confused. They know it is their bedroom, but something else tells them it is not, as if some vital connections are not being made. The visual cortex is saying, 'This is your room', but the emotional connection, or connections of memory, are casting this into doubt. It reminds me of Capgras syndrome.

As Carla tells me, 'I think part of it is also these experiences remind us that we are so close to madness at all times. I think maybe that is what is at the core of the fear. The subjectivity of experience, and the fear that you might not be able to turn it off. And the knowledge that no one knows what happens to you in certain states. No one knows what happens after you have died, and there are these big mysteries in life... Those mysteries would just be filled with this kind of madness, and darkness and consciousness. The other thing with sleep paralysis is that part of the real world is part of the fantastical world.'

Scientists may be able to tell you, 'It's the amygdala', or point to other parts of the brain as explanation, but what may lead to the big production numbers of your sleeping imagination doesn't explain the cast. 'I don't really trust a huge amount of my perception of so-called reality, but you are just meant to live with that,' Clara concludes.

Be sceptical of your reality, but try to avoid it teetering too far into paranoia, is how I attempt to live my life. The insecurity that is created by subjectivity should not be overwhelming... too often.

'Material existence is entirely founded on a phantom realm of mind, whose nature and geography are unexplored' – Alan Moore

Writer and artist Alan Moore has thought a great deal about reality and has placed us in numerous versions of it, through his many stories. They are a way of changing a reality that would be far more stubborn to change if you attempted to physically alter the world. He can build towns through his imagination rather than through concrete, taking us to the edges of the universe without building a rocket. The mind is a good place to travel in.

Alan is also a lucid dreamer. In his dreams, if he decides that around the next corner there is going to be a fantastic mansion that is his to live in, then he'll walk around the corner and there will be that mansion. It saves years of construction work. 'You apply your will to something that you've imagined and eventually you can materialize it. In the world of dreams that can be done instantly, whereas in the world of data that will take you decades, potentially. So it seems to me that maybe they're on a continuum, these two worlds: the world of our imagination and the material world.'

Sometimes a conceptual existence can seem almost more powerful than a physical existence, and reality may be a negotiation involving emotional, psychological and intellectual reactions. As Alan says, 'Everything is mind – at least for us. Is the thought of a unicorn a real thought? And how does that change the reality of the unicorn? The problems come when we demand that others believe in or obey a reality that is very personal; when arrogance enters the frame, that demands all others banish their subjective perceptions because you decide your perception is the ultimate truth – whether that is with a vote count in Pennsylvania or the behaviour of your gyroscope. Once you are demanding your perception to be taken seriously and in a way that will affect those around you, then evidence should be required. It is good to make yourself familiar with the imagination of your mind and its predilection for duplicity and self-deception.' Alan believes that spirit of his intellectual life to be in a constant state of questioning and trying to arrive at answers that seem to satisfy the individual, while hopefully being stringent enough in his reasoning to not make too many glaring errors. It is a consensus reality, but he worries that we don't all live in it. It's nominally a consensus reality.

At a book-signing at the Hay Festival, a mother told me that her daughter had a question. The daughter took a deep breath and, with a vague grimace, asked, 'How do I know that I am not merely a creation in someone else's dream?' And so as we snaked along in the queue, we tried to work out a system for challenging this possibility. At ten years old, this girl was asking the same question Descartes had tried to answer. We set up a series of experiments, but I don't think we came across a solution that was conclusive. At times, we came close to a simulation theory – an infinite regress of dreams, where each different world is a dream from another world. The girl may be in someone else's dream, but what if her dreams created within this dream are further worlds? And can we ever find our way to the original dreamer?* Sometimes I like to imagine that my dreams are my

---

* The actual science of dreams has become very interesting in the last few years and has journeyed beyond Freud, to testable ideas of their purpose beyond paranoia that you have incestuous desires. It seems that REM sleep and your nightmares help you deal with trauma. After a nightmare, it appears that should you then recount a genuine traumatic experience you have dealt with, the physiological effects of recounting the trauma are reduced. This is connected to the extreme reduction of the production of noradrenaline during REM sleep. People with PTSD continue to produce noradrenaline during REM sleep, and this leads to a reduction in the effect of trauma after a nightmare.

brief, fevered visits to the other worlds in which the splinters of me exist. Sure, it is bullshit, but bullshit can be fun as long as you don't stake your life on it.

I have always had a penchant for ghostly stories. The fantasy of the chattering of the ethereal but persistent dead may seem a polar opposite, but there is something in this strangeness of unlikely events that seems to have a similarity to the possible physics of the black hole and other similarly counter-instinctual phenomena. It is an eeriness. In writer and cultural theorist Mark Fisher's *The Weird and the Eerie* he writes, 'The perspective of the eerie can give us access to forces which govern mundane reality but which are ordinarily obscured, just as it can give us access to spaces beyond mundane reality altogether. It is this release from the mundane, this escape from what is ordinarily taken for reality, which goes some way to account for the peculiar appeal that the eerie possesses.'

I am not sure what Professor Brian Cox would make of me placing theoretical physics next on the shelf to the *Dark Tales* of Shirley Jackson and H. P. Lovecraft's *At the Mountains of Madness*. He might have quashed the existence of ghosts with his understanding of the second law of thermodynamics, but I find that a haunting eeriness is given breath with this theory of the holographic universe. There is the sense that we are not where we are now, firm and solid, but that we are two-dimensional projections from somewhere else in space. It can make me feel as if I am a ghost already, even before I die. Here's to the new ghosts, the illusions that we may be, and the reality that entails. Keep chipping away at the walls of your reality tunnel, but don't jump to any conclusions – solid conclusions can be where the danger lies.

The best advice I have read is for us to take our reality seriously, but not to take it literally.

The truth is a lemon meringue.

FRIDAY O'LEARY IN *MR GUM*

CHAPTER 10

Imagining There's No Heaven – On Being Finite

The sudden chill where lovers doubt their immortality

As the clouds cover the sky, the evening ends

Elvis Costello, 'Couldn't Call It Unexpected No.4'

The size of the universe, the evolution of living things, the nature of spacetime – these are all things that started to be revealed by human curiosity. We never needed telescopes to reveal death, but our increased knowledge of the physical world has certainly changed our expectations of it. It can take a bit of getting used to.

One of the longest-running conversations in *The Infinite Monkey Cage* series has revolved around how to pinpoint when a strawberry is actually dead. I won't go through it all again; suffice to say that, despite hearty debate – and even suggestions that the possibility of life coming from the seeds in a pot of jam may mean that a strawberry has eternal life usurping any notion of the sell-by date on the base of the jar – we were eventually given a clear answer by a botanist. Our strawberry was buried with pomp, ceremony and cream.

Science has not yet been able to give a definitive answer for when life on Earth began. The earliest traces of life have been found in sedimentary rock dating back at least 3.5 billion years, with some likely to be as far back as 3.8 billion years, but the point where this planet went from lifeless to living is as elusive as finding a satisfactory definition of life itself. The moment this planet is a dead place again may be a little easier to pinpoint.

Professor Steve Jones defines life as 'Entropy held at bay by extracting energy from the environment'. His definition of death, though, is even briefer: 'entropy wins'. By the look of things, entropy always wins. One day you, I and everyone we know in the universe will stop extracting energy from the environment, and that second law of thermodynamics that keeps popping up will have had its way. It will be a sad day. It will be an inevitable day.

I was always a morbid brat, so I am afraid it was inevitable that we would find ourselves facing death somewhere near the end of this book. With our increasing understanding of what life consists of, it becomes more and more difficult to believe that we get to do anything more after our life on Earth comes to an end. It is harder and harder to avoid the feeling that this is finite. You will meet fewer scientists in this chapter, just in case you were hoping that I would offer revelations concerning how we will live for eternity in the form of a microchip or one of those holograms I keep harping on about.

Research into the afterlife has not been at the forefront of scientific investigation lately. Many people now hold a healthy doubt that they will ever experience a full and conscious afterlife with an intact self. Such doubt stems from many sources, including the reduced need to posit the existence of God in order for the universe to exist, and an increasing understanding that the soul of us is material, rather than something separate and beyond or above the physical stuff that makes us.

Paranormal investigator Hayley Stevens told me about the work of Dr Sam Parnia, who has carried out research with cardiac-arrest patients, examining their experiences of what happens to us as we seem to be on the edge of life and death. Some of his research has included placing suspended symbols on boards facing the ceiling in resuscitation rooms, to see if those who have been near death, without brain or heart activity for a period of time, have had an out-of-body experience that has seen them float up and be able to read these symbols. The results have been negative so far.

For the time being, I am living my life as if this is the only go I get. Though medicine, science and technology will give many of us longer lives, there are limitations, and death remains inescapable – however many implants, replacements, nips, tucks and total transfusions may become possible. Archaeological evidence, from the Druids of Stonehenge to the pharoahs of ancient Egypt, suggests that human beings have always felt the need to fill in the mysteries of death with stories about the afterlife. With no get-out clause on offer from science, I wonder how we can find ways to cope with death without recourse to probable fictions?

One of the morbid hero-fixations of my childhood was Peter Cushing, the elegant portrayer of vampire-slayers and obsessive scientists. Cushing was introduced to death by his mother when he was little. If he was naughty, his mother would punish him by pretending to be dead. He found this greatly upsetting. She would start to sing a song about drifting over the sea and into death. He would promise to be good, but she would punish him by dying and then sitting stonily in an armchair. His brother would berate him for being so upset. 'Kick her. Shove her,' he would say, but Peter was unable to. One day he was particularly upset at his mother being dead again and he happened to have a slice of bread and jam in his hand, so he shoved it in her face. Cushing said she learnt her lesson and never did it again.[1]

Death changed for me, after my son's birth. It's a common parental occurrence. It came into focus a few weeks afterwards when I visited an exhibition by Walter Schels and Beate Lakota at the Wellcome Collection. Schels and Lakota had interviewed dying people and took portrait photographs of them shortly before they died, then shortly afterwards.[2] I was unable to look at the portraits of children – something I know I could have done before the birth of my own son, however uncomfortably. Now I had family ahead of me, as well as behind me.

This changed my relationship with morbidity. My own death was now a pragmatic issue; I must not die, as I had someone to help to adulthood. My fear of death was replaced by a fear of loss. I think my son was nearly ten years old before I stopped standing outside his door at night to check he was breathing. If you ever want to see relativity in action, experience how long a second is when you are waiting to hear your child's next breath.

When I return to the house that I grew up in, my first thoughts are often about birth and death. At night I sleep in the same room that I was born in. A February snowstorm made a hospital birth impossible, so at 6.30 a.m. on 20 February 1969 Dr Webber arrived at the edge of the village with a shovel and dug his way through the snow and to my crowning head.

In the room next door to this, my mother died forty-six years later. It was a much calmer winter night. The nurse didn't need a shovel to get there. We had spent a night and a day listening to my mother's increasingly hollow breathing. Sitting on the bed opposite, with a classical music radio station playing lightly in the background, we would lean in every now and again, listening out for her elusive breaths. The transition between life and death seemed so slight. Until that point we could still imagine that there would be a sudden rallying and a demand for a cup of tea and a bun. It was what happened with my great-aunt, May, who had been a nurse in the First World War and lived well into her nineties, a resilient battler. Shortly after she was declared dead, she woke up, asked what all the fuss was about and why her son was crying. She had a brandy and her life went on for a few weeks more.

When a family member dies, there is the leaving of a presence; it feels like more than a physical departure. You visit their home and something has gone. It is not like they have gone out to the shop – there is a presence that is not there any more. I am fond of the proximity to birth and death in my childhood home. It provides a frame and a story – a tangible attachment to entrances and exits, which is useful when your beliefs are rooted in a physical and finite existence.

Although I believe that death is death, and that is that, I have fun imagining improbable possibilities. I imagine that when we die, our consciousness comes loose from our skulls. Our thoughts drift out into the air around us. For a short time, without any sensual experience we realize that we are still alive only as thoughts. There is no looking back down on ourselves or hearing the words or the weeping. Just as this ethereal me starts to dwell on this afterlife of consciousness, a breeze begins to disentangle me and I quickly thin until I am no more – bits of consciousness spread across a meadow and then further apart. Billions of years in the future, the ethereal consciousness finds itself brought together again for a fraction of a second.

I now realize that this whimsical daydream has a pleasing connection with an idea known as the Boltzmann brain. In billions of years, when the structure of the universe has succumbed to the second law of thermodynamics – which predicts that every system becomes increasingly disordered, even a system as big as a universe – at that point there will be no more stars or planets or life... most of the time. In the vast fug of particles, due to the enormous amount of time involved, every now and again things will come into existence. It is predicted that one of those things would be a brain. It would be a brain with memories of the past and a view to the future; a brain that would not exist for long, but would believe that it had a life. It might even believe that it has written a book or that it is reading a book *right now*. And then it would be gone. Some scientists consider this more probable than your or my belief that we are in reality (there's that word again), living a life now on a planet. If you don't really believe you have a life at all, except as a brief pattern of particles experiencing a grand illusion, then it certainly gets rid of having to worry about the afterlife.

Anil Seth has experienced death in life, and you might have done too. He has reached the point of non-experience that is as close as you get to death without being defined as death. 'Just over a year ago, for the third time in my life, I ceased to exist' is how he opens his TED talk. He explains that his brain was filling with anaesthetic. After feeling an increased sense of disorientation and detachment, he ceased to exist, then a little later he was back. Unlike sleep, where there is a sense of then and now between falling asleep and waking, Anil had no sense of whether five minutes, five years or fifty years had passed. He simply wasn't there. He describes anaesthesia as turning people into objects and then, hopefully, back again into people. During his period of non-existence Anil did not get a day-pass to any afterlife.

Disproval of an afterlife by scientific research isn't top of the list of things that will get a research grant, but it seems a hard picture to maintain, with current views of the universe. If we are to presume that death destroys us, then perhaps we have to work even harder to establish how we accept it during life.

Where are the funeral photographers?

Callum Cooper is a psychologist who describes belief in the after-life as a win–neutral situation that no one should be afraid of. You either die and are reunited with those you love, or you die and there's nothing. If you do hedge your bets on everlasting life, I think it is still best to live your life as if this is all there is. It saddens me to think of all the nuns and monks living their itchy lives in hessian and rough cotton, expecting Pearly Gates and juicy grapes in the next life, and then it all being for nothing. Mind you, they also mastered very strong tonic wine, so that heavenly inebriation probably helped them get through some of the lingering days of calligraphy and scratching.

The real problem is not the death itself, but how we live knowing there is an end, and how we continue to live when people we love die. Evidence-based thinking can be wary of ceremony and ritual, but maybe we need it even more when a funeral is not an act of sending someone elsewhere, but an acknowledgement of a life and a mind that are no more.

Callum has a particular professional interest in bereavement and death. He believes the English have an unhealthy relationship in dealing with death: that it is too distant, something to be hidden, sealed shut as quickly as possible. He saw something very different while travelling in Egypt. A dead body was being carried through the streets. Apparently a man had died by the side of the road and so people picked him up and started to carry him. Some would carry him a few steps before passing him on, others would continue further with him. This was crowd-surfing of the dead. Each person helped so that they could feel they had been part of his journey. They had connected with death.

Callum says that for us a lot of death is more a representation than a reality. You assume your loved one is inside the coffin, but you don't get that hard-hitting reality of 'There is the cadaver.' That face you saw laughing and telling tales is now without potential. They have lived and now they have died. Their body was the shell that made them get around, made them think, enjoy everything, eat and drink, and argue and fall out of friendship and fall back in again. Callum believes that by evading this connection, we have a skewed sense of death and loss. Our detached way of dealing with death may make us more susceptible to increased death anxiety. He gets feedback from his students about how they were first exposed to what they would consider 'unusual' funerals, such as ones with open coffins that might start the day in the front room of a house, for all to see. Some of the students would be particularly surprised when mourners would line up to kiss the deceased on the forehead. Once you are the dead, for a few cultures all connection ceases.

At the funeral of Callum's grandmother, he asked the undertaker to take some photographs of him standing by her coffin. The undertaker was briefly unsure, but then happily obliged. For most ceremonies marking the landmarks of life, the cameras are out; but more often than not, there will be no photographs of our final show.

'Let's read *Peanuts* at dusk'

For some people, the thought that this is it is truly unbearable. I have lived a fortunate life so far, in that my losses have been few. For those who have lost young people to war or nature, it may be harder to dismiss the idea that this life on Earth is all there is. My mother believed in an afterlife. She had lost her father when she was a child and she was determined to see him again. I would not have wanted to persuade her otherwise. I argued with an atheist friend who insisted that it would be wrong to pretend to anyone that you believed in the afterlife if you did not (we had been drinking). My feeling was that the advantage of not believing in a god gave you one less being that you had to obey. Odin, Zeus or

Shango could not strike me down for my white lie about heavenly belief. One night when my mother was anxious and delirious, she became increasingly upset because 'scientists had proved there is no heaven'. I found no conflict of values in saying that scientists had done no such thing – it was not one of the issues they were working on at CERN.

I am not a comedian who deliberately sets out to shock audiences, but the subject of death has caused the angriest audience reaction I have experienced over the last decade. It was while I was performing with Brian Cox for what was probably the seventy-first or seventy-second time that we had done the show. There was much in the show on the understanding of time, from the perspective of physics. As usual, most of my contributions were to lighten the load after some of the more complex ideas had been delivered by Brian.

Near the end of the show, I would talk about our psychological battle with the passing of time and then recite a poem I had written about a day of playing with my son, making a den in the woods. It was in the same woodland that I played in when I was a child, not far from the house where I was born and where my mother died. As we played and talked, I became very aware that the time we would build dens together might be nearing its end, as he was nearly ten years old. After this, the only time we would build shelters from sticks and wood would be if we ended up in a post-apocalyptic world where we were forced to forage fungus, eat shrews and fight off others with fragile bows. After my son fell asleep that night, I wrote a poem about the day. I don't often write poems, but that night one came to me. This became the poem that I would recite each night near the end of the show. Here it is:

You don't need a storyteller now
Your bedtime is almost autonomous
But still
One snuggly hug
For safety from the sandman
Is today the day?
Is this our final den?
We dragged the sticks and rolled the logs
Made jokes about those passing walkers
With those weird-shaped dogs
You found our furniture
A worn and mossy tyre
And I warned you of all the dangers
Of that leaf-hidden rusty, rusty barbed wire
And then
Damp-bottomed we sat
And viewed our architectural feat
I phone-filmed your pride
For the archives of things we've done
The woodland adventure
Of father and son
Sometimes
Walking hand-in-hand
I secretly mourn for the days that are not yet gone
The days that seem like a Shepard sketch for an A. A. Milne
Where every beach is a post-war postcard
The blue, too blue in my recall.
Your freedom is necessity
But not yet

Not yet

Just wait a little bit

Let's pond-dip for skaters with a net

Let's build a sofa train

A Lego piece found

By my BARE FOOT!

Let's read Peanuts *at dusk*

And Calvin and Hobbes

Let's dig and splash and play

And mime laser-deaths in outer space

LET'S RACE!

And then I'll let you go

But not yet

Not yet...

One more day?

After I performed the poem, I would walk offstage and Brian would return to the spotlight. He would connect my memory of that den-day to the laws of physics he had discussed. He explained that the very laws that mandate there will be life in the universe are the same laws that mandate there must be death. He would focus on that realization: that we only have so much time, and that should inspire us to live well and kindly, with curiosity, trying not to waste our days, but seeking out beauty whenever we could; that we must savour the magical days. The spotlights would go out and then George Harrison's 'All Things Must Pass' would play as Brian and I came back to take our bows.

On this particular night, so many months into the tour, as we drove to our hotel I received a very angry message on social media. I take upsetting people very personally. Although this was only one of 5,000 people, that did not change how important their reaction was to me. The message said that it was a 'shite ending' and the writer wanted to know all the details of the producers and promoters, to lodge a complaint. I tried to engage with them. Our seeming irresponsibility had been to connect the beauty of the universe – and the laws that create that beauty – with the inevitability of death. It transpired that the fury came from a mother who had lost a child. The family had come for a fun night out and ended the evening being unavoidably confronted by their grief. I tried to explain that the poem and the show's conclusion came from a place of love, not antagonism or nihilism. I tried to explain that the poem, at least to me, is not about death, but about the passing of time and how our awareness of it hopefully means that we might savour what we have.

I was told the only solution was to excise the ending in its entirety, because I had clearly never thought how many people in the room might have lost a child. Eventually I gave up attempting to explain our intentions. This parent had lost a child and her reaction was real, and I saw no satisfactory solution. Sometimes, when talking to people who have suffered greatly, you need to accept their feelings, but also accept that the true cause of their suffering and anger was not you. It was a stark reminder of fragility and loss, and the terrible possibilities of life.

I do not know if that person had religion in her life. I do not know if it was the poem, or the moment when Brian said that the laws mandating there must be life also mandate there must be death, that caused her distress – or, more probably, both. That night I thought again of the John Updike quotation,

'Astronomy is what we have now instead of theology. The terrors are less, but the comforts are nil.' Much as I have fought against the notion that the comforts of science are nil, did this event, and that mother's reaction, really prove that Updike was right all along?

I received one more angry message, and then the communications stopped. I still had things I wanted to say; I still felt the need to explain myself. Though she might have thought that each night I had unthinkingly walked onstage oblivious to the fact that some in the audience might have experienced great loss and sadness, that would not be true. The very first night I performed the poem, in a small room in Northampton, I was aware there was someone in that room who had lost their son in an accident.

'I can't believe I had to kill my own son to get you to visit...'

I first met John Ottaway after he tweeted that he would have liked to come to a show I was doing, but he was currently between jobs and strapped for cash. I hadn't used my allocation of complimentary seats, so I told him he could have them. When I returned to that venue the next year, I remembered John's name, partly because it is similar to that of John Otway, an eccentric performer and 'two-hit wonder' with whom I had once worked.* I sent John Ottaway a message via social media offering him free tickets again. He replied, 'Sorry, I can't – my son has just

* John Otway had a post-punk Top Thirty hit, 'Really Free', in 1977, and in 2002 his song 'Bunsen Burner' reached number nine. He is a highly energetic performer and while on the TV music show *The Old Grey Whistle Test*, a particularly vigorous leap led to him injuring his testicles. This TV moment was turned into a flick-book to promote his movie *Rock & Roll's Greatest Failure*.

been killed in a motorbike accident.' Knowing that he was an ex-soldier, I thought this was some sort of dark joke, or maybe he had hated the last show so much that he had to create a cast-iron alibi. But it was true. On the last day of his apprenticeship, John's son Jamie was killed in a road accident.

I got to know John a little better over the coming weeks. He came to another gig a few months after Jamie's death. He had a quandary. The year before, he and Jamie had been at the Cheltenham Science Festival. They both had a good time there and were planning to return. A particularly fond memory was of Jamie persuading his dad to approach his hero, Alan Moore. It was one of those bonding moments, like me building a den with my son. John still wanted them both to go back, but Jamie would return in an urn. John asked me if he thought it would be okay if he brought Jamie's urn, dressed in a T-shirt, and asked the scientists they had seen the year before to pose with him. I was pretty sure it would be fine. I told the scientists about John and Jamie, and they were all willing. Without exception, everyone said yes to the photograph.

John, Jamie, his mother and his sister were not religious. When it came to life after death, Jamie's interest was more closely connected to George A. Romero's zombie movies, such as *Night of the Living Dead*, rather than the mystical trappings of Jesus and Mary. Jamie might be dead, and the rest of the family might not have the consolation of believing he is alive somewhere else where they will eventually re-join him, but they have found ways to be with him, through events and ritual. Their communion is not red wine, but a fruit-juiced-based soft drink, J2O. They pour it down the drain nearest the scene of his death because of its connections to Jamie and his life – when he was a teen, Jamie had got 'drunk' on J2O, not realizing it was non-alcoholic.

The lack of religion has not reduced their connection to their son, and it might even have made it stronger. The lack of the comfort of faith has been replaced by the need to create events, to keep Jamie part of their existence in the place where he can live: their memories. So although he is not there, they keep creating new memories that involve him. This includes regularly putting his ashes inside fireworks and shooting them into the sky. They celebrate his life more than they mourn his death. John says it is not about them, 'It is about Jamie.'

For John, too, there is comfort in the scientific story of our ever-recycling atoms. He can look at the grass in a few years' time and think that some of Jamie's atoms are part of that grass now; and, in a typically John piece of logic, 'The bird that shits on me – some of that shit will be Jamie.'

For those without religion, the story of recycled atoms offers the solace of continued connection. It was the story that I told my son when he was seven and started to worry about death. Around that age, many children first experience an existential terror – that moment when they start to worry, 'Mum and Dad might not be here for ever... and I might not be, either.' Not having recourse to the afterlife can make things difficult in many ways, and I couldn't say to my son, 'Don't worry, we all go to heaven.' The first thing I told him was that anyone who said they knew what happened after we die was talking rubbish. Then I told him the atom story. It was Carl Sagan who introduced me to the idea that we were all star-stuff: 'The stuff of us is the stuff of the stars.'

I told my son about his atoms being created in stellar nurseries, and of their journey through the universe. Some may have been in the rings of Saturn, some may have been in the meteorite that struck the planet Earth and led to the end of the dinosaurs. Some may have been spewed out of volcanoes, some in waterfalls and kangaroos and lemurs' tails and great white sharks. For now, they are my son's atoms. But there will come a time when those atoms begin to drift off and eventually they will find themselves having even more adventures: in mountains and whirlpools, in minnows and giraffes, in apples and walruses, and eventually beyond the planet Earth; and one day they will be in a star. Sure, I was a little loose on specifics, and I may well have anthropomorphized the atoms, but the basics were true. No equations were either harmed or involved, and it was enough of a story that it stuck.

A few days later, my son returned from school with some worries about his atoms. 'Dad, you know I'm going to become a star?'

'Yes?' I replied worriedly, wondering how fast and loose I would need to be with my malleable story.

'I just wanted to check: it is only going to be *my* atoms in that star, isn't it, and not anyone else's?'

And here I lied. 'Oh yes, it will only be your atoms.'

'Oh, good, as there are a couple of boys at school that I don't like very much, and I was worried I was going to be stuck with them for billions of years.'

So the star-stuff afterlife story is far from watertight, but I still think it's a good story, and it is one that grows with curiosity and doesn't stay the same, with the same answers and the same presumed reality.

Old MacDonald had a graveyard

Some of us spend more time than others imagining our funeral. We see people being turned away from the packed venue; the tears; those more successful than us praising our influence and admitting they would be nothing without us. Then we imagine the Jay Gatsby funeral: the rain, and an owl-eyed man being one of the only mourners. We renew our gym membership, hoping to put off our depressingly sparse funeral for as long as possible.

Shari Forbes's interest in death began when she was sixteen. She started reading a lot of crime novels, which led to her studying forensic science at university. She saw an autopsy as part of her degree and was fascinated. At twenty-one, she knew she wanted to study death, and particularly decomposition. She wasn't nervous about seeing a dead body. She was fascinated not merely by the autopsy, but also by her reaction to it. She can look at a body, disassociate from any emotional response and examine it from a point of scientific fascination. To her, the body is just the shell, housing someone who has now left. When her grandmother died, her mother wanted Shari to have a moment to talk to her. Shari tried to say something emotional to the dead body, but it didn't feel natural because it wasn't her grandmother.

I first saw Shari giving a very persuasive talk in Toronto on why leaving your body to a body-farm is a good post-death ambition. Body-farms are where corpses are placed in different environments to monitor and record the process of decomposition. When people first find out that Shari works at a body-farm they ask things like 'Doesn't it smell?', 'How do you cope?' and 'Where do I sign, to donate my body?' Many people are fascinated and may consider donating their own bodies, but among those who don't understand it all, Shari has found it is usually for religious or cultural reasons.

She used to be very scared about death and the end of the world. It was her greatest fear as a child and was perhaps what drew her to those crime novels, but now Shari doesn't even think about it. Having seen the realities of what dead bodies are, her fear of death has waned, and she believes that we can reduce such anxiety for many people by having more exposure to death and by talking about it more. Shari likes the idea of a lifecycle and of cycling. She sees some of the process, as she observes the decomposition, of what grows from and around the bodies. Things keep cycling; you die and you decompose, and you go back into the earth in some way. Everything needs to keep cycling for this world to keep turning. This is her afterlife.

She has a very physical view of how, as we decompose, we become habitats and new useful molecules. She sees mushrooms, fungus and insect mounds that suddenly form right next to the decomposing body. They are all using the resource of the body to cycle through their own lives, and then they will do the same thing again and again; and as long as there is life, it will be an ongoing process. 'I love it when we can actually see these physical realities of decomposition, because that's how it works' is the way Shari sees it.

Is there meaning in becoming mulch, to hatch out from? When you come to believe that your body is an empty and detached thing after death, and all the *you*-ness of you has been vaporized, some people find meaning in the use of the husk that has housed them. Although we will not be there, our attachment to our bodies in life means that we may plan to give them purpose while we still can, even if that means donating them to a body-farm. It is another story to help us through life.

Bill Hamilton was an evolutionary biologist. His work had a great influence on Richard Dawkins's *The Selfish Gene*. His pioneering work was on the genetic basis for altruism. His altruistic thoughts are apparent in his funeral plans:

> I will leave a sum in my last will for my body to be carried to Brazil and to these forests. It will be laid out in a manner secure against the possums and the vultures, just as we make our chickens secure; and this great *Coprophanaeus* beetle will bury me. They will enter, will bury, will live on my flesh; and in the shape of their children and mine, I will escape death. No worm for me, nor sordid fly, I will buzz in the dusk like a huge bumble bee. I will be many, buzz even as a swarm of motorbikes, be borne, body by flying body, out into the Brazilian wilderness beneath the stars, lofted under those beautiful and un-fused elytra which we will all hold over our backs. So finally I too will shine like a violet ground beetle under a stone.[3]

Not everything requires a testable hypothesis

Zena Birch is a humanist celebrant. She lost a lot of people close to her when she was young. When Zena was eleven, her eight-year-old brother was run over. It made no sense. Finding better ways to acknowledge the death of friends and family, and the connections with those left behind, is important to her.

Humanist weddings still have no legal status in England, but as funerals have no legal status anyway, humanist ones are as much funerals as any other. After all, however much you might wish it, a funeral cannot be annulled. The status of 'dead' is hard to revoke by paperwork alone. Zena stresses the importance of ritual, of feasting and partying, but mainly of annotating time in a different way. Of the first humanist funeral she attended, she says, 'I was absolutely blown away by it being the most appropriate starting point for grief that I've ever really been part of, because the ceremony was so based upon that person's life, and their interactions with each of the people in the room.'

We need a ritual for loss, even when we are without faith – perhaps even more so then. Church funerals can seem bureaucratic. They have a structure that frames people who may well be falling apart. There is a time slot and you'd better not overrun, because there is another one at 3.20. But grief cannot be scheduled.

When Zena's brother died, she remembered people saying that he'd gone to a better place, or that God wanted him back, or that God took the best ones – and she felt lied to. It did not seem to express a reality. She was a child but, like most eleven-year-olds, she understood far more than others acknowledged. Her suspicion was that these were platitudes with little thought behind them; the structure of it enabled people to avoid any deep interaction with others' loss – 'poor things', 'God will take care of them' – and move on. The difficult conversation could be avoided; and what a difficult conversation it is, in English culture. We may be more comfortable crying due to losing a football match than losing a sister.

During the funeral, as Zena heard people say her brother was with God and with Jesus, she thought, 'I don't even know who these people are.' She was certain her brother would rather be with her and their mother. All she saw was devastation. This was the beginning of her struggle with religion. No one was telling the truth, because the truth didn't seem acceptable or right.

Zena explains that the core of what she does, as a humanist celebrant, concerns how we engage with story – not fictional story, but real-life story – and how that enables us to understand our own feelings and empathize with other people. This is the core of why she likes doing what she does. She believes there is often a need to be irreverent, because death is not reverential. Death makes total sense in terms of the laws of physics and its mandates, and the necessities of biology; it just makes no sense to us that people we love, people who were vibrant with existence and experience, simply stop. The physical needs to keep their consciousness alive cease to fulfil their requirements, and the person we joked with yesterday is utterly non-existent tomorrow. The absurdity comes from being something that knows it exists, and something that is aware that this situation will not be for ever. Once we accept that there is an end of consciousness, and we find it impossible to correlate an afterlife of consciousness with what we have learnt, then we either evade the inevitable until entirely necessary, or we have to think even more deeply about how we face up to inevitable loss and find comfort and meaning in the brevity of it all. Although those most rigidly demanding an evidence-based worldview with *no exceptions* may blanch, I can understand how death can create little cracks in our logic, and in those little cracks we might find a connection that can bring some comfort.

Zena's mother sought many ways to make sense of her son's death. The only one that gave her at least a little comfort came from an Indian spiritualist she met in Exeter. He told her that all children who died transform into butterflies. If she ever saw a butterfly, then she knew that was Phil giving her a little nod. The physical reality of butterfly possession is of no concern; the logic of butterfly incarnation is unimportant. The story is beautiful, and its reality is not necessary to create a moment of connection and of living on through memory. Here is a brief burst of benevolent cognitive dissonance.

Zena told me about a Peruvian fire ceremony that had been passed on to her by someone who had met her globe-trotting friend Laura. You take however many years of a person's life and you start a fire. Once the fire has started, there is a log put on by a different person for each year of the dead person's life. Before they put the log on the fire, they get to share a story or say a few words and then the log gets added. Laura was thirty-seven, so there were thirty-seven stories. It started with her mum and dad, then went through her friends. 'It went from the sublime to the utterly ridiculous and from the rude to the perfect,' recalls Zena. As the stories increased, so the fire got bigger and bigger, and by the end everyone was warmed by this giant physical fire made from memories and love for Laura. The fire brought out the candid nature of storytelling, but also real grief. Zena saw that some people were totally capable of telling a brilliant story, while others could barely make it through a sentence or two. In the light of the fire, you could see everybody's different relationship to grief. Zena believed it freed everyone up in the way they were allowed to feel about Laura and the loss of her. The experience was visceral. Let me put into print now the request that I have a Peruvian fire ceremony when I am gone, and you are allowed to bring marshmallows (but no sausages). The one thing that can live on after we die is our stories, so it is worth trying to make as many as you can.

'Something enormous had happened and yet nothing had happened'

I have had far longer conversations about ageing and death with biologists than with physicists. Obviously ageing and death are in the realm of biology, whereas physicists are usually thinking on a much bigger or much smaller scale than mere pesky life. But some of the most touching thoughts on death that I have heard come from a physicist.

The theoretical physicist Richard Feynman's first wife, Arlene, died when they were both very young. The story of their relationship is told in his essay 'What Do You Care What Other People Think?' It is a story of love, and then a story of battling with denial of the truth, and finally of a mind being scientifically curious during death while not detaching from the psychological experience of it, though he tries. When Arlene first fell ill, the doctors seemed baffled and Feynman educated himself at the library, and from what he gathered, the prognosis was not good. When it seemed almost irrefutable that Arlene's condition was terminal, Feynman was told to keep it secret from her. This troubled him greatly, as their relationship had been based on openness and honesty. 'I wrote Arlene a goodbye love letter, figuring that if she ever found out the truth after I had told her it was glandular fever, we would be through. I carried the letter with me all the time.'

Arlene's persistence and demand for honesty eventually led to Feynman telling her that she had Hodgkin's disease (it was later diagnosed as tuberculosis of the lymphatic gland). The essay becomes a story of love in the shadow of inevitable death, and Arlene is a figure of tremendous wit and resilience. On her last day Feynman talks of listening to her breaths and thinking about what was going on physiologically. Losing the person he loves most in the world, he sits with his curiosity about why her life is fading. After she died, he leant over to kiss her and smelt her hair. He was surprised that it smelt just as it had earlier when she was alive, and it made him think that 'something enormous had happened and yet nothing had happened'. This is the confusion, this can be the fury, this is the calamity – that everything looks almost exactly the same and yet nothing will ever be the same again.

Why the hell wouldn't we invent stories with futures for the dead? Of all the denialism I have mentioned, I think death-denialism is the most understandable. Feynman believed he was fighting against the reality. After she died, Arlene would turn up in his dreams and when he would try and tell her she was dead, she would explain that she was only pretending; she had got bored, so she cooked up a ruse, but now she liked him again, so she had come back. A month or so after Arlene's death he saw a dress in a shop window and thought, 'Arlene would like that', and it was in that moment that it hit him. Arlene was gone – there would be no return, save for the occasional comfort or disturbance of a dream.

When Richard Feynman himself was dying, documentary film-maker Christopher Sykes and his wife, Lotte, made a film about his interest in the Tanna Tuva People's Republic, a socialist republic that existed on the Russian Mongolian border in the 1920s and 1930s, before being subsumed into Soviet Russia. Christopher had made the superlative Feynman documentary *The Pleasure of Finding Things Out*, which was predominantly of the physicist seated in an armchair, talking of his life and work. There was also a spin-off series of short films, *Fun to Imagine*, in which Feynman explained jiggling atoms, rubber bands and big numbers. The two of them had built up a relationship, and I would particularly recommend that you seek out the episode where they argue over the explanation of magnets.

By the time they filmed *The Quest for Tanna Tuva*, Feynman was gravely ill with abdominal cancer, something he had lived with for the last ten years of his life. One morning, after much debate, Christopher decided to ask Feynman whether he would talk on camera about his thoughts on death. Feynman was just getting up as Christopher went in to ask. He remembers that Feynman was putting on his socks as he said to him, 'I want to ask you if you'd mind if I asked you about the fact that you're dying. Obviously if you don't want to talk about, that's fine. But if you did, that would be fine, too.'

Feynman got back into bed and said he would think about it aloud, because he really didn't know whether to talk about it or not. And then he started talking, and he told Christopher all about Arlene and about how, after her diagnosis, they had decided to live what remained of her life as best they could. Then he talked about imminent death and mortality. After all this, he came to a conclusion. 'You know what? I find it very depressing. I don't think I do want to talk about it.'

Feynman's friend Danny Hillis, an engineer and author, amongst other things, was walking in the hills with him shortly after he had had another major operation for his cancer. Feynman was telling funny stories about befuddling his doctors, but Hillis was unable to hide the gloom he felt that his friend was dying. Feynman asked him what the matter was.

'I'm sad because you're going to die.'

Feynman replied that it bugged him too. 'But not so much as you think. When you get as old as I am, you start to realize that you've told most of the good stuff you know to other people anyway.' Of all the scientists who lived in the twentieth century, it seems that few have left so many stories that are so often told. When I look at the shelf where all the books by and about Richard Feynman sit, I then look at the other books above and below – each one filled with so many stories, some fact and some fiction, but all created by human minds, each one leaving behind something of what they were and what they dreamt. There are footprints and fossils of memories on every bookshelf and in every library. Write your story down; leave your tracks on a page for someone to find.

Death is what makes life absurd. It's why Samuel Beckett has all the best lines on it. 'They give birth astride of a grave, the light gleams an instant, then it's night once more.'[4]

It is the time limit that gives us our drive. Would you get anything done if you knew you had an eternity to do it? I bet heaven is a place of good intentions and very little happening. Anyway, eternity is too long. I hope for a long life. I hope that the curiosity of other human beings keeps working out ways of giving as many people as possible long lives and good lives. I have a lot of books I still want to read, but I think the spring of eternal youth may have a downside. You'll have to watch the planet die, and then the universe die, and you'll have a long wait before those blue whales pop in and out of existence again.

The only universe we will ever know is the universe with us in it. We can imagine before us and we can imagine after us, but in that imagining we are still in the picture: an observer in a bubble protecting us from the material possibilities of what we see. We cannot imagine the world without us, but there will be one. The universe of your experience, though, dies when you die – that universe shuts down as you do.

The transition from being alive to being dead needs to be problematic. It must be worthy of contemplation or we are not acknowledging how remarkable conscious life is. How strange to be aware, when almost everything we see around us isn't. How fortunate we were. This should drive human creativity to find as many ways as possible to prevent people being robbed of life too young. The ambition for longer good lives is fuelled by the realization that this is probably it. Death reminds me to try and treat people well now, and not to spend too much time being frustrated by petty quibbles and squabbles. I try to tell people the positive things I think of them now, rather than wait to compliment them when they are dead in a box.

After my son was born I found myself worrying about death pragmatically, thinking of what might happen if he became fatherless. As I would cross a busy road, I would think, 'You'd better take more care now. You have read enough psychology to know that a child who loses a parent can have a whole new box of strife and confusion to deal with, consciously and unconsciously.'

For everyone who is a support for others, who is a connection, a place of love or kindness, a companion, it is of little consequence that on a cosmological scale it is all pointless. It is on the small scale of terrestrial concerns that there comes meaning. There is a tag match, that in our life we hope to infect others with 'the better angels of our nature', hope that we can leave behind seeds of happiness, that we can jemmy open the potential to have a good life, to diminish suffering, to leave footprints of kindness. There are daily eulogies to be found on social media; in between the duelling and the spats and the spite, you will find fond memories of people gone.

On the day I am writing this, I am looking at people's shared memories of Georgge Jeffrie, a comedian turned writer, I worked with many times when I was young. He has unexpectedly died. Phrases like 'sheer kindness, fun and decency' keep being repeated. He has left behind his influence – not some grand influence that transforms art, cures scabies or changes the architectural design of a foyer, but an individual influence to all those who walked into a room and thought, 'Ah, good, he is here. This will be a good day.' Nothing refines a sense of purpose more than knowing that you have so little time to have it and use it. There are those days when you find yourself returning home and thinking, 'Today I think I created some happiness' or 'Today I think I left someone feeling less lost than when the day began.'

The rest of the universe will know nothing about this. It will not alter the speed of entropy or change the shape of a galaxy, and the only evidence of its existence may be in the immeasurable thought of one other person, but that's enough. Committing yourself to love means committing yourself to loss. At times it cannot seem right that burning brightly comes with knowing the flame will die. I've got this far through the book without quoting Buddha, but I think he had a good point when he said, 'All composite things decay. Strive diligently.'

In the end, eternal life would erase any need for meaning; purpose comes with the limitations of existence. It's a drag, but whoever the hell thought it up was right – all the good stuff comes from the end always being near.

All was forgiven.

All living things were brothers, and all dead things were even more so.

KURT VONNEGUT, *THE SIRENS OF TITAN*

CHAPTER 11

More Important than Knowledge – On the Necessity of Imagination

I'm still that kid sitting alone in the middle of the night,
thrilled just to look through the window at my piece of the
universe, wondering what else is out there.

Janna Levin

The stairwell of Huddersfield Library is my favourite stairwell in England and there's some stiff competition, as anyone who has toured the stairwells of England will know only too well. Not only does it lead to an art gallery, but it is daubed with graffiti quoting from some of the finest minds the world has ever known. It was here that I was first confronted by Albert Einstein's statement 'Logic will get you from A to B. Imagination will take you everywhere.'

I have been lucky. I was lured back to science whilst browsing around a bookshop in a town steeped in witchcraft. Like many of us, one of the reasons I turned away from science was that it did not feel like a place of the imagination. One of the reasons I wrote this book was because I have met so many people who have been under the misapprehension that scientists are Vulcans. While artists are seen as dynamic eccentrics filled with joie de vivre, booze and magic mushrooms, scientists measure angles and keep dull-coloured coded notebooks of things they have counted; and yet from my experience, the hotel bars at science festivals are more joyously chaotic than any book festivals, and the ideas fly.

We can feel that the arts are fields to run wild in, while science is a cage where you are either right or wrong. The stories are often replaced with equations to memorize, things to know, but not necessarily to understand. It comes as a surprise to some people that the ability to recite an equation does not necessarily mean that you understand the universe, though frequently governmental education advisors do not seem to know this. Talking to science teachers now, they remain frustrated that those in power who make the bold curriculum choices still fail to see that there are stories to be told.

'I'm a massive number of atoms put together in a particular combination, and that's okay'

Alom Shaha is a science teacher. When he was a child, his main interest was in the arts and writing stories but, fortunately, despite excelling at this, he was offered very little encouragement, whereas his science teachers were keen to cheerlead him through experiments, whether he failed or succeeded. Early on, his science teachers explained to him that he too was a chemical reaction and was not separate from the physical world explored in science. This is an important part of learning about science. There is not you and then science on the outside, but you are part of the process – you carry many of the stories of the universe with you and inside you.

He found out that science was not abstract. We can never truly be an outside observer of nature. We are in the thick of it. We are a zoo, both outside and in: your skin is crawling with eight-legged microscopic mites, with the microorganisms inside us outnumbering our human cells by ten to one. Are those voices in your head really you overhearing the mites? Are you beginning to itch now? Each square millimetre of your skin is a Pixar movie just waiting to happen.

Alom thought the place for stories and imagination was the English class, but then he discovered that science didn't mean leaving the world of a well-told tale. In education and beyond, the importance of science is often stressed as giving us the power to do things, to innovate technologically, to try and control nature, to eradicate disease, to travel beyond the planet. All these things are true. We will be relying on the scientific imagination to tackle some of the greatest problems that we will be facing over this century. But this view makes science utilitarian. It gives it purpose, yes, but if we only see science as a solution, we rob it of its delight and romance. We put a wall around it.

Rather than give children what is required to function in society, science should be equipping children with the desire to explore and question, to interrogate the world and the universe, allowing them to leave school with the tools of critical thinking that will make them questioning and self-questioning. As the anthropologist Margaret Mead wrote, 'Children must be taught how to think, not what to think.'[1]

Few people leave school without knowing some stories about how the universe began – whether it was being born from a giant egg or fashioned from darkness and void by an omnipotent and lonely god. Despite this, many leave school without being taught the story of how it appears the universe actually began (obviously missing out the very first bit, because that initial ten to the minus forty-three seconds or so is proving tricky). Gravity and quantum mechanics don't go well together, it seems.* We should know where we came from; and not just biologically, via broom-handle demonstrations. All these stories – the mythic, religious and scientific – come from the human imagination. All required a 'what if' and some form of internal logic, however flamboyant, but the scientific answers had to be testable, too. It wasn't enough to be a good story, though it *is* a good story. It may lack serpents or copiously vomiting deities, but it can still make an audience gasp.

I have sat in audiences and heard physicists explain the expansion of the universe, or they have talked of a time when everything was the size of a grapefruit and yet it was also infinite, and eyes pop out and there is an astonished, almost silent mumble of incomprehension. It is a moment of extreme and startling wonder, and it is not merely the children reacting; it is people of all ages. For me, every grapefruit is now cosmological. As I go to slice one in half, the daunting thought of the entire universe fitting in my hand can overwhelm me. It's true of oranges, too, and if I am not careful I'm aware that this sort of knowledge could lead to scurvy. It seems a travesty that so many

* More of that later; for now it can be a cosmological cliffhanger.

of us might get through life without having some knowledge of those moments of the universe that seem so utterly absurd when you first hear of them.

How strange that the most remarkable and most likely story of the beginning of everything is not given the same position of significance as the myths. It doesn't mean we must dismiss the myths that came before, as they are part of the story. Alom feels that part of the problem in science education is that it lacks the creativity offered by the arts. Students who do English, music or art at school are encouraged to produce original work; they may well experience the thrill of being a writer or a musician or an artist. But how do we create the experience and the thrill of being a scientist? An inspiring education system should demonstrate that science is a creative endeavour.

Alom was fortunate in that he was encouraged to create his own experiments, to test himself and the laws of the universe. He decided to try and check something called the Mpemba effect, named after the Tanzanian schoolboy Erasto Bartholomeo Mpemba, who discovered that warm ice-cream mixture froze quicker than cold ice-cream mixture. Alom tried to investigate this with basic school apparatus. He used his scientific knowledge and intuition to devise the experiment and investigate it himself. He describes it as 'absolutely disastrous. My results were totally bonkers and made no sense.' Although the results didn't show anything, he got a good mark because he investigated every failure scientifically – because he was trying to problem-solve. This was an experience that excited him and engaged him, even if it was a failure. This is another important lesson: that failure is necessary for progress.

Watch a professional juggler and you may well see that when it comes to the biggest trick, the first attempt at juggling a chair, a kumquat and a leopard does not go to plan. This may well be a deliberate error. The suspense increases, and the jeopardy makes us lean in further: is this trick even possible? Failure is part of the process. Failure is often a good story. The most exciting stories in stand-up comedy do not involve a comedian telling you how they went onstage and everyone loved them; they are about being run out of town by an angry mob after the leopard escaped and ate the club chairman, whose escape was foiled by slipping on a kumquat. But the story is made all the better if you know that this person is also a success. They are not always run out of town, but that doesn't mean they are a stranger to failure.

Science is not about coming up with an idea, testing it and confirming that it is right; it is also a story of many exciting ideas that got nowhere. Many of the 'what ifs?' lead to 'Ah, that was not what I thought would happen at all.' Many 'maybes' become 'maybe nots'. It may be frustrating, but it is also thrilling. This is a story filled with suspense. The first Hollywood feature film about an *Apollo* mission was not about *Apollo 8* successfully orbiting the Moon, or *Apollo 11* leaving footsteps on the Moon; it was about *Apollo 13*, a mission that failed in its objective, but where, against the odds, the three astronauts made it back to Earth alive. It is a very successful story of a failure.

The errors, missteps and feuds that lead to great leaps forward and sometimes stumbles into ditches help show that science is a human project, and that the passions and vanities of humans play their part. Good ideas replace bad ideas, and then better ideas come along. The answers are often still good and useful and have a purpose, it is just that there are places still to travel. Science is not stationary.

The beautiful Palace of Science

When Andrea Wulf won the Royal Society Science Book Prize for her book *The Invention of Nature*, she said that she couldn't wait to tell her childhood chemistry teacher. This was not because her teacher would be proud of her, but because that teacher had written her off as being totally useless at science. Wulf is one of many children who were told, explicitly or implicitly, that they didn't have what it takes to 'do science'. From that point on, she felt shut out of that world. *The Invention of Nature: The Adventures of Alexander von Humboldt* was her way back in. Fortunately, despite being told that she did not have the kind of mind required, her curiosity was persistent. Andrea said she thought science was like 'a beautiful palace, with many, many doors, but we only show kids one door to go through. We can't all go through that door. I was lucky, I found another door to go into, which was biography. It's still the same palace.'

Anne Goldsworthy became an expert on primary-school science education mainly because no one else wanted to do it. She now runs an independent primary science advisory service. The reason no other teacher wanted to take on science was because many found it scary themselves. Many had hated science when they were at school and suddenly they were being asked to teach it, and they had a fear of getting things wrong all over again. Anne had not enjoyed science at school, either, but here was her chance to find a way to repair what she considered to be the mistakes of her teachers. She had found science boring and disliked the fact that it took no account of her ideas. Things were either right or wrong and there was no room for play. Science learning in her childhood had avoided the journey of how we test what we believe, and find out why things are as they seem to be or why they may not be quite as we imagined.

Anne believes that science coincides well with the way children learn precisely because there is play in it, and that nothing else on the curriculum offers quite such opportunities for exploration. Your personal thoughts, opinions and ideas are always relevant. She shows me a series of drawings by children on how we digest food. It is a smorgasbord of messy and delightful ideas. It started with a discussion about what seems to be happening when we eat: food goes into our bodies, makes a journey and comes out somewhat differently, certainly less deliciously. But the children were asked, 'What goes on in the middle?'

Gemma's drawing had the food going down the throat into the stomach to get crushed. Lucy's expanded on this, with the food not merely being crushed, but carefully moulded into shapes that are easy to fall out of our bottoms. Another child had all sorts of little men inside him; some men take the food to the heart to pump blood, some men throw the waste food down the hatch; other men painted the food brown and gave it 'a bottomy smell'. The right or wrong answer is not important. It is about thinking how the body works. It is about possibilities. It is the adventure of thinking and drawing.

Then came the experiments, such as squeezing gloop through old scratchy tights to see how the useful ingredients of food make their way through the body, while the waste makes its way to the bottom. After all this, they discussed what they had found out. Without shame, they often said, 'Well, I thought it was like this, but now I know it is like that', in a way that would be far harder for most adults to admit. They were given the tools of experimentation and supposition that lead to finding out about the world, in a way that memorizing twenty-two equations doesn't give you. Anne would happily banish exams in exchange for actual learning of *how* and *why*, but the education system seems to require results that can be catalogued, rather than being happy with the broader results of creating intrigued and interested young humans.

Anne shows me another series of tests about how we see. Most children initially believe in emission theory, so they are with Plato and Empedocles on that. Many children imagine that their eyes fire out beams that illuminate objects. With a simple experiment using a cardboard tube, they tested this and found that if you cut out a light source, your eye does not illuminate what you are looking at. 'It doesn't work like the way I thought it did, it works another way.' 'This was amazing.' 'I felt baffled – I am baffled. I was confused all day, but then...' This is science as joy, not merely as numbers and symbols. Play first, equations later. There is a story in every earthworm, cloud and fingernail; it's just waiting for you to find it. As I once thought W. B. Yeats wrote, 'Education is not the filling of a pail, but the lighting of a fire.'[2]

For many scientists, their first childhood inspiration came not from the classroom, but from reading science fiction. Here were worlds of the imagination unrestricted by the shackles of contemporary technology, able to play fast and loose with the laws of physics: the possibility of alien encounters, unlikely evolutions, non-material consciousness and multiple universes. The emotional attachment to *Doctor Who* and Philip K. Dick is often still strong in grown-up scientists. In his *Hitchhiker's Guide to the Galaxy* books, Douglas Adams introduced me to many scientific ideas; it's just that I didn't know that they were scientific ideas at the time – I thought they were pleasing, absurd nonsense from an excellent comic mind. The line between contemporary physics and absurdist comedy can be thin at times. Dead cats that are alive – or is that living cats that are dead? – are simply the tip of an infinite universe of the almost unfathomably weird. Except it is not even weird, it is exactly as a universe should be, simply not what we expect it to be.

While science fiction is rich terrain for stretching the imagination, it is odd that science fact can be viewed as a sterile room that asks you to leave any fancifulness of the mind in a bucket at the door. Mathematician Jacob Bronowski created one of the greatest television series about science, the landmark series *The Ascent of Man*. Many books of his articles and lectures were published in his lifetime, and his celebration of scientific endeavour was often underpinned with passionate explanations of the need for the human imagination, if we are to understand the universe. He wrote, 'We do a great harm to children in their education when we accustom them to separate reason from imagination, simply for the convenience of the school timetable.' It is also notable that his first two books were about poetry, not science. In *The Visionary Eye* Bronowski wrote, 'Many people believe that reasoning, and therefore science, is a different activity from imagining' and he then proceeded to strip apart this idea with zeal.

It would be easy to blame the poets, but they have probably suffered enough, as poets so often do. John Keats famously bemoaned that the poetry of the rainbow had been disenchanted by Newton turning it into a prism. In 'Lamia' he wrote of science* that it:

Will clip an Angel's wings
Conquer all mysteries by rule and line
Empty the haunted air, and gnoméd mine –
Unweave a rainbow.

Bronowski took the world's most famous equation, $E = MC^2$, and adapted it into poetry using one of Keats's most famous works. In 'Ode on a Grecian Urn' Keats wrote:

Beauty is truth, truth beauty, – that is all
Ye know on Earth, and all ye need to know.

Bronowski changed it to:

Mass is Energy. Energy Mass – that is all
Ye know on Earth, all ye need to know.[3]

Has he made a lovely thing dull, or has he taken an equation and shown the beauty of it by placing it in poetic form? What beauty is there in knowing that all solid objects – all blackberries and raspberries, volcanoes and elm trees, possums and babies

---

* To be clear, he actually writes 'Cold philosophy', but at that time science was known as natural philosophy and I think the investigation he is talking about is what we would now define as science.

– are energy? Every lush landscape gains an explosive potential. Holding a tomato in your hand, you can contemplate the forces within that maintain its delicious integrity.

Mars or Magritte?

Physicist Tom McLeish works at York University, where they have re-created the chair of natural philosophy just for him. This allows him to flit from physics to anthropology to medieval studies. On art and science, he quotes Karl Popper: 'A great work of music, like a great scientific theory, is a cosmos imposed on chaos – in its tensions and harmonies inexhaustible even for its creator.'[4]

Tom thinks that science and poetry can be driven by similar desires, and that our reaction to them can also share similarities. We don't have to be an expert to enjoy either. It is a part of being human to wonder at the world. When he first saw a Rothko painting at Tate Modern, he could not see the point of it at all. He had his children with him, and they too seemed quite angry that something like this was hanging in a gallery. Now he would happily stand in front of a Rothko for half an hour or more. He has found something fulfilling in what was incomprehensible to him. Sometimes, if you keep looking at what you do not understand, you may experience a eureka moment, and suddenly everything fits into place. You have been building up the pieces in your mind, repeatedly trying to piece them together, but it has been like taking a hammer to a 10,000-piece jigsaw puzzle of St Paul's Cathedral. But then one day – without hammer in hand, without consciously moving the pieces – things fall into place. Everything fits and you see the bigger picture.

In his book *Helgoland*, Carlo Rovelli explores what an understanding of quantum physics can mean to our reality. He writes about superpositions. This is the situation where, on a quantum level, something exists in two contradictory positions, in two places at once. It is a concept that is naturally distressing to consider, because it is not how we see reality at our level of experience. It is not how we believe things are meant to be. It contradicts all of our experience. Befuddlement is too quaint a word to summarize the spasms of my mind that I experience when I try and get my head around it.

As I read, shushing the voice that screams, 'You don't understand this! Give up! Give up! Give up!', I come to a paragraph where Carlo Rovelli writes, 'We never see a quantum superposition. What we see are the consequences of the quantum superposition. These consequences are called "quantum interference". It is the interference that we see, not the superposition.' I had read these words before, in similar contexts, but on this occasion, as I sat in a hot bath rereading and rereading, something made sense to me that I had not seen before – I had a new image in my head. Sometimes it can be big moments of clarification, but very often it is just another baby step. What was confused or alienating seems to reassemble in your mind, and where once there was confusion, there is now something that is transcendent, and it usually catches you unawares.

In her book *The Canon: The Beautiful Basics of Science*, Natalie Angier wrote of our changing attitudes to art and science. The parent takes the child to the science museum, but one day the science museum is placed at the back of the wardrobe with our dolls and teddy bears, and we are taken to ponder the Picassos and say something that we picked up from a Sunday newspaper article about the nature of the brushstrokes or the use of light, and pass off this critique as our own.

If only our museums were not divided by discipline. You immediately notice the change in atmosphere by the change in volume: a science museum can have an echoey cacophony like a municipal swimming pool, whilst the art museum is as silent as a members-only library for nonagenarian monks. I like those municipal museums built by nineteenth-century Unitarian tea importers, where woodland animal taxidermy butts heads with the painting of a Pre-Raphaelite nymph, all in the shadow of a recently restored steam plough. I am a particular fan of improvised taxidermy, where you see how a nineteenth-century taxidermist who has never viewed a kangaroo has had to take whatever skin and bones have arrived in the post and work out roughly what shape it must be. The Natural History Museum in Dublin has an over-inflated orang-utan that I am specially fond of.

I would like to be the curator of such a museum and would call it the Museum of All Curiosity. My Rothko would be next to a projection of *Voyager*'s pale-blue dot, and the Giacometti bronze statue would be reaching out for the Crick, Watson and Franklin double-helix model. This would make it harder to set our mind to a default position for the environment. Go to the National Gallery and we have our 'I'm looking at art' head on – the head that tilts at a slight angle and aims for an expression of deep and mature fascination. In the Science Museum we have our 'I am trying to understand cellular division' head on – the one that leans into the contraption, smiles and cranks the handle of the one demonstration machine that doesn't have an 'Out of order' sign on it, until the attendant comes over, asks you to cease cranking and takes out another 'Out of order' sign.

Does my brain know when I am looking at art and when I am looking at science? Does it monitor my reaction, when I am beguiled by an image of the surface of Mars, differently from the feeling that occurs when enraptured by a Magritte painting? Both move and intrigue me, but do they move and intrigue me in different ways? Is there a discernible difference in my delight? Does my mind separate them into their disciplines? Does nurture dig a ditch between the two? The great divide is going from an environment of 'Touch the things and push the buttons' to '*Do not touch*' because your sweaty fingers will taint the beauty.

The art museum reminds us that we are a corrosive presence. Science is for contact; art is for distant gaze. Science is enhanced by your contact; art is stained by it. My brain struggled with the York Art Gallery when I found that various statues and carvings said: 'Please touch'. I could only believe that these signs had surely been placed by some prankster, or a Yorkshire cop short on busts who decided he could meet his yearly quota by framing a naive southerner, caught red-handed stroking the nose of a Jacob Epstein. What next: a hand-crank on the side of Rodin's *The Kiss*? A pop-up tab on the *Mona Lisa* that you could pull, to make her smile? When I finally overcame my paranoia and touched the nose of the Epstein work, I was relieved to leave the gallery uncuffed, but my anxiety was so great I decided that I would nevertheless avoid touching art again.

Brian Eno's work has straddled art and science. He has been ingenious in his use of technology to create soundscapes for moonscapes and exhibitions of conceptual art. He is a thoughtful artist engaged with science. Despite this, when I asked him about the conflation of art and science, he told me that he has come to believe that it is a bad idea. From his experience, he has seen that when art and science try to create an event together, it leads to 'very bad art and very disappointing science'. He thinks this is because they don't do the same thing.

Brian has a simple four-word phrase, which he knows is a simplification of what science is, but it helps him think about it: 'Science discovers and art digests'. For him, science is constantly finding out things about the world and saying, 'Hey, it seems to work like this.' It is the way we understand how to do things, and look after ourselves and so on, but the knowledge itself doesn't tell us anything about how we should use that knowledge. He doesn't believe that science exists to give us these kinds of answers; instead, for him, this is where art comes in. 'Art says to us: "Look, this is where we now are, and how do you feel about it?" Novels, for instance, can say, "This is the state that we are in, how do you feel about that?"' Brian sees that art can be the digestion process of scientific understanding.

He tells me that 'The industrial revolution was the discovery of a whole lot of new ways of controlling nature, new ways of using energy, new ways of creating tools, new things that tools could be created for... but none of that in any way tells you that you are going to end up with a completely disenfranchised working class, living in the shit in Manchester and dying of lung disease at a very, very early age. What tells you about that is Charles Dickens, or writers of that kind, who say: "You know what, this is the world. This is where we are. What do you think about that?" That's the digestion process that art does, I think.

And, you know, it's clear we need both things, but don't get them mixed up.'

Maybe I have been lucky or maybe I have been naive, but I think I have seen art and science working together well. Maybe it is my cognitive dissonance kicking in; after all, I have spent a lot of the last fifteen years creating situations where art and science mix, and I would hate to think those nights where I brought together The Cure with some epidemiologists and an astronaut had been a waste of time. At the very least, we discovered that the unruly hair gene of Robert Smith was a gene that made Dr Ben Goldacre's hair unruly too. Obviously the effects of rubbing your hand repeatedly through your hair as you compose lyrics about existential anxiety does the same work as rubbing your hand through your hair while attempting to find a statistical pattern.

It is possible to interweave science and art in an engaging and creative way. 'Cape Farewell' was founded in 2001 by the artist David Buckland. Its purpose was to find new ways of introducing people to the effects of climate change. It is committed to bringing together artistic and scientific projects. It has taken artists on expeditions to areas of the world affected by climate change. Scientists explain what is going on and then the artists go away and create work as a reaction to it, whether through music, painting, sculpture or jokes. At one event, I saw beautiful, tactile and intricate sculptures of molluscs. Then a marine biologist explained how the shells of these creatures were rapidly changing, due to increased acidification of the oceans.

On just such a trip, the artist and bookseller Natalie Kay Thatcher made some beautiful art comic books inspired by

American theoretical physicist Richard Feynman,[5] which eventually led to her putting on the exhibition *Jiggling Atoms*. Here was artwork that came from artists' reactions to the ideas of quantum mechanics and other theories of physics. After wandering around the sculptures and paintings, they could settle down on flip-up chairs and listen to discussions about supercolliders. How can we turn science into art, or art into science, is the not the same as how can we understand how imagination leads to science and art, and where these ambitions overlap.

We have a way to understand how to do things and look after ourselves, but that knowledge itself doesn't tell us anything about how we should use that knowledge. I find the border between how we gain knowledge and how we use it a little blurry. Sometimes I think scientific knowledge carries with it a strong idea of how it changes us, our meaning and the purpose of it.

When I ask Tom McLeish about Brian Eno's position on the psychological effect of art versus science, he is careful to preface his response with a reminder that experience is subjective. He wonders if, when Brian Eno is looking at an image such as *Apollo 8*'s *Earthrise* photograph, he experiences a sense of the technical achievement, whereas when he sees a Raphael he might go into an apoplexy of artistic excitement?

What does art mean? It has much to do with what you bring to the art or the creation or the visual image. It is not all about us, and neither is it all about the artist. Can a scientific image or idea, when understood with the appropriate back-story, give rise to the same emotions of rightness or glory or insight or awareness that you may experience with art? Tom believes '100 per cent, yes.' What is true of both art and science is that a little knowledge adds to the beauty. To know about the processes of J. M. W. Turner – his mixing of colours, his use of beeswax to create depth in the paint, even his short-sightedness – makes the story more fabulously intricate, just as the landscape of Mars becomes richer as you discover more about how its craters and mountains were formed, and how human beings developed the technological expertise to be able to take a photograph of the planet's surface.

Wandering around the Walker Art Gallery in Liverpool, I stopped at a Turner painting of dusk and I stared at it for so long that the jovial attendant began to wonder if I was casing the joint. I was thinking of the processes that made the painting – not only the processes of the artist, but the processes of nature that Turner had interpreted on his canvas. The nuclear fusion that created the light that then travelled towards the Earth, some of it striking the ground and then being reflected and reaching the rods and cones of Turner's eyes, firing his imagination. His brain had evolved to a complexity that meant he wanted to communicate his subjective response on canvas. Every part of that process could be explained with equations and diagrams except that final part – the imagination and consciousness that created the image, the workings of a mind that is still with us today, long after the artist died.

The visionary artist Cecil Collins believed that artists enabled people to experience what they couldn't see themselves, and that is true of the scientist, too. Van Gogh has enhanced my view of the night sky, but so has solar physicist Joan Feynman. Among her many achievements was working out how many high-energy particles were likely to hit a spacecraft in its lifetime, and understanding what causes the aurora borealis. When she was young, her brother Richard took her to a golf course where they watched an aurora. Spellbound by the lights, Richard explained that no one really knew what caused them, and this set Joan off on an adventure of discovery. Richard was older than Joan and was already making waves in the world of physics, and she realized that she did not want to compete. They made a deal on dividing up their areas of investigation. She told him, 'I'll take auroras, and you take the rest of the universe. And he said OK!' Richard did okay with the rest of the universe, sharing the Nobel Prize in Physics with Sin-Itiro Tomonaga and Julian Schwinger, for his work in quantum electrodynamics. He once said, 'Poets say science takes away from the beauty of the stars, mere globs of gas atoms... What men are poets who can speak of Jupiter as if he were a man, but if he is an immense spinning sphere of methane and ammonia must be silent?'[6] But perhaps Richard had not read enough poetry.

Edgar Allan Poe may have called science 'the vulture of the heart', but other poets found inspiration in this evidence-based inquisitiveness. Samuel Taylor Coleridge was a friend of the chemist Humphrey Davy. Coleridge encouraged Davy in his poetry, and Davy welcomed Coleridge into the laboratory. Is the combat that we imagine between poets and scientists more territorial than ideological? If science can offer beauty, does the poet fear redundancy? Is it better to reject science rather than see that it offers new pictures of reality to play with? Do some scientists fear that a glaze of poetic mysticism will get in the way of objective discovery?

For a scientist who could also be very disparaging of philosophy, Richard Feynman was a frequent contributor to the philosophy of curiosity.* In *The Pleasure of Finding Things Out*, Feynman tells the story of an argument with an artist friend about the beauty of a flower. The artist, like Keats, considers that the more you understand the flower, the less beautiful it becomes. Feynman then runs through what we now know about a flower, beyond its pretty petals. The molecules, the atoms and aesthetics. 'The fact that the colors in the flower evolved in order to attract insects to pollinate it is interesting; it means that insects can see the color. It adds a question: does this aesthetic sense also exist in the lower forms? Why is it aesthetic?' He concludes, 'Science knowledge only adds to the excitement, the mystery and the awe of a flower. It only adds. I don't understand how it subtracts.'

Knowing that each part of the story of a flower and its development has a function does not turn it into something that is purely functional. It adds layers; it transforms it into multitudes. Though the status of a story may change, from fact to fiction or to myth, the story is not lost. The storybook is even fuller after each dissection.

The great neurologist Oliver Sacks, who did so much to popularize ideas of the layers and complexity of our brains, especially when in turmoil, wrote of his burgeoning fascination with flowers in his essay 'Darwin and the Meaning of Flowers'. The garden

---

* 'In the area of philosophy of science, though, like many physicists of his and the subsequent generation (and unlike those belonging to the previous one, including Albert Einstein and Niels Bohr), Feynman didn't really shine – to put it mildly,' wrote professor of philosophy Massimo Pigliucci, which is why I have been a bit looser and just called it 'the philosophy of curiosity', which seems general enough that I might get away with it; https://aeon.co/ideas/richard-feynman-was-wrong-about-beauty-and-truth-in-science

of his childhood had many flowers, as well as two magnolia trees, 'with huge but pale and scentless flowers'. Noticing that the ripe magnolia flowers attracted little beetles rather than bees and butterflies, Sacks wondered why. His mother explained that magnolias were more ancient on the evolutionary tree scale and so existed before more modern insects such as bees, and so more ancient insects were needed for pollination.

'Bees and butterflies, flowers with colors and scents, were not preordained, waiting in the wings—and they might never have appeared. They would develop together, in infinitesimal stages, over millions of years. The idea of a world without bees or butterflies, without scent or color, affected me with a sense of awe,'[7] wrote Sacks, and with those words he infects us all with that sense of awe when we wander around a garden, inciting us to tilt our heads towards the petal trumpets to see how evolution led to working relationships between plant and insect. The view of a pretty garden is made more vivid by our minds playing with the events that occurred in each plant's origin story; the shapes and colours are made richer by a sense of purpose. So knowledge does not always add beauty, but it can increase our fascination with the world, and it often gives a beauty or a charm to things that have perhaps seemed ugly or discomforting.

Tom McLeish told me two stories of differing passions that dispel that contrast between the traditional cold demeanour of the scientist, frigidly acknowledging that their evidence-based investigations have been accurate, versus the hot-blooded artist whose temperament boils over with effusive passion. Tom was at Durham University on the day of the announcement of the detection of gravitational waves, and he burst into tears and hugged his friend,

David Wilkinson, an astrophysicist and theologian. Tom was overwhelmed. Here was the detection of something that was smaller than the quark in a proton atom. He felt a sense of comradeship towards all those scientists who had gone before, working on something considered to be near-impossible, an important proof of the ideas of Einstein, but something that Einstein thought would never be detected. This was work by people who knew they might provide a rung on the ladder, but they would not be the generation that would see it come to fruition; and Tom was part of the generation that, thanks to their generosity and foresight, would see it.

One of Tom's Durham compatriots was Carlos Frenk. He is the sort of cosmologist who gets positively excited by the idea of being wrong. Tom's other story is about when Carlos's life's work on cold dark matter came into question, due to observations of satellite galaxies made by a Durham research student. Carlos came bounding around the corridors, beaming. Tom wondered what he was excited about – after all, this discovery could undermine his life's work – but Carlos's reaction was, 'Yes, isn't it great!' He was excited by the possibility of new knowledge, rather than anxious about his own work.

Carlos's area of study was Lambda cold dark matter and the spectrum of the thermal correlations of the cosmic microwave background radiation. This is the radiation left over from the Big Bang, and the observation of Lambda CDM is a vital piece in understanding the story of the Big Bang. The tiny permutations of heat shown in this famous image predict where stuff is in the universe.

The predicted thermal correlations were on a curve, with four or five wiggles on it. When the data from the COBE satellite came in, it showed the incredible accuracy of the predictions. Tom explains that such precision in prediction is rare. There were all the points that scientific imagination believed should be there – bang, bang, bang, bang, bang – spot on. Tom says that despite science giving the impression it is always like that, it almost never happens that way. When Carlos gives talks and he gets to that moment about the accuracy of the predictions, Tom always sees him choke up. The emotion of imagination and technology coming together to create such accuracy is inescapable. Carlos says he's good at hiding it; that he coughs, turns away and clears his throat, but Tom knows he's choking up, even though he's told this story many times. For him, this moment is as beautiful as any symphony. This is not Vulcan behaviour. These moments are important; they are connections between the adventure to unravel the universe and the emotion when each new outcrop of information is discovered – even if, as in Carlos's case, it goes against all that he had held true throughout his working life.

I have spoken to Carlos on many occasions. He is a highly intelligent man, who often looks at me as if he's amused by some eccentric animal that has been pickled in a jar. I think we are equally perplexed by how the other's mind works. After recording a show about cosmology with him during the Edinburgh Festival, he came along to see one of my stand-up shows. It was one of those shows where my mind went into paroxysms, making absurd rambling connections to neuroscience with the sort of delivery that, without microphone and spotlight, would see you get free entry to the nearest asylum. Carlos's mind finds order in the universe; mine creates mess.

Carlos has a great love for the arts. As we talk, if he isn't telling me about the Einsteinian view of time or the problems of uniting gravity and quantum behaviour, he's explaining why I must watch Ingmar Bergman's film of *The Magic Flute* (I did and I was glad). He finds strong connections between art and science in many paintings. He is a great admirer of Magritte. When he gives public lectures, he uses paintings to make connections to cosmology. He starts by explaining that we don't know what most of the universe is made of, that we only know a tiny little sliver. He explains that great big parts of the universe are dark energy and dark matter, and we don't know what they are. He explains how he found that it was natural to represent this with Magritte's painting *The Lovers II*.

The dark energy of Surrealism suggested by
Magritte's *The Lovers II*.

'It's kind of natural to put those two things there to represent the dark matter and dark energy. They may be related. They have a veil. So they don't know about one another. And, in fact, you cannot even see their faces. So to me, that was a perfectly reasonable way to try to portray to the public dark matter and dark energy,' Carlos explained.

When he talks about the origin of galaxies, he may refer to Michelangelo's *The Creation of Adam* in the Sistine Chapel. He may end his lecture with *The Garden of Earthly Delights*, focusing on the Earth element, though hell represents the dark energy because that is where Carlos believes it must come from. As far as he can tell, the universe should not have dark energy. He feels betrayed by whatever, or whoever, thought dark energy was a good idea, and believes that if there was an agent that created the crazy laws of physics, then it was almost certainly a mischievous one.

'Did you ask a good question today?'

Lucie Green is likely to have a far greater appreciation of a Turner dusk than I am. She is a solar scientist. Her first observation of the Sun through a solar telescope was, for her, a eureka moment – her moment of realizing that the Sun is so visual. This may seem obvious; after all, the Sun illuminates our world, but what we see on a beautiful summer's day is far from the true magnificence of the Sun's activity.

Generally, for most modern human beings, the Sun is merely a bright disc in the sky that we know we are not meant to stare at, especially during eclipses. With a solar telescope, Lucie saw the

The Sun (but you knew that).

detail of so many different features. The Sun felt so layered for the first time, and she was drawn in. Take a look at NASA's thirty-minute film *Thermonuclear Art* and you will start to get some idea of the perpetual spectacle that makes life on this planet possible.

Lucie had not initially gone into science as a career, because she thought of it as a pursuit lacking in passion. She wanted to commit herself to something that she felt could be expressive and passionate. She feared that science would force her to be a mere data-processing robot. When she finally specialized in solar physics, she felt there was a strong sense of constraint, of putting a lid on her character. She told me, 'If you were lively, passionate – you know, maybe you just saw something that was so amazing that you shouted it out loud – then that was a bad thing.' In one of her first appraisals she was called 'ditzy'. But twenty years on from that experience, she believes the situation has changed, and that those who wanted her to suppress her delight and wonder belong to another generation and are no longer the dominant voice.

We talk about how vital it is to engage people in science as a human endeavour, often driven by emotion; and of how emotion plays a part in *why* we want to know what we want to know – that graphs alone will not win someone round to your opinion. For Lucie, her experience of the Sun has changed her relationship with it. 'Studying the Sun means that now I know it – I know its character. I know what it's doing all the time. I can check out any minute of the day to see what it's up to, because we've got the kit now that enables us to do that. So I feel I have a companion all the time, whereas previously it didn't even register.'

Lucie's passion for the Sun reminds me of my first visit to the Tate Modern. The gallery itself is a brick-and-mortar collision of art and science. What now houses Hockney paintings and Barbara Hepworth sculptures used to be home to two pairs of 25-kW Brush arc-lighters and two 100-kW single-phase alternators generating at 2 kV and 100 Hz. This was a power station that fed the street lamps of the City of London and Southwark. Now it feeds the artistic curiosity of school groups asking, 'But is it really art, Miss?' as they look at, say, Joseph Beuys's *Lightning with Stag in Its Glare*. Although the Turbine Hall was amply filled when this was a powerhouse, it is easy for art to shrivel and sulk in such a vast space. Not so for *The Weather Project*. The power station was dominated by a nuclear reactor: the Sun.

This was a spectacular artwork by Olafur Eliasson and was described as 'the basis for exploring ideas about experience, mediation and representation'. It was sun worship with the benefit of a roof, just in case it became drizzly due to the sun god feeling let down by less-than-wholehearted worship. It created a hazy dusk that had people basking for hours under its luminescent light. When you left the gallery your sense of the Sun, and all the vitality it generates, was changed. This was the kind of art that deeply affects an audience, because after experiencing it, your world is changed. I have met artists with Messiah complexes, but here was an artist benevolently playing God, or at least one of the gods – perhaps the sun god Helios.

Olafur also created installations for the UN Climate Change Conference in Paris in 2015. He has been described as bottling lightning, 'so that we may appreciate its delicate majesty'.[8] His art is often based on his fear that we are obsessed with the present moment and forget about the future.

His work is art, but there is science in it, too.

As well as writing one of my favourite science books, *How the Universe Got Its Spots*, Janna Levin wrote the novel *A Madman Dreams of Turing Machines*. As a child she never identified as a scientist, partly because of the way she felt science was portrayed. It seemed to be 'This is that and this is that, and this is how it is.' She couldn't believe that it might be creative or interesting or exciting. She now sees it as something more nuanced. 'It's not a list of facts. It is also not 100 per cent a list of unknowns. We know a lot. We do make discoveries that are concrete that we can move from, that are like jet fuel. We're somewhere in between these two ridiculous extremes of "everything is just pretend" and "everything is a list of facts".'

Janna worries that the problem for many people is that science has been portrayed as lacking ambiguity. One of her favourite stories is about Professor Isidor Isaac Rabi who won the Nobel Prize in Physics in 1944. He wrote, 'My mother made me a scientist without ever intending to. Every other Jewish mother in Brooklyn would ask her child after school, "So? Did you learn anything today?" But not my mother. "Izzy," she would say, "did you ask a good question today?" That difference – asking good questions – made me become a scientist.' Janna feels that to be questioning is the most important scientific impulse, and that science is therefore a process, not a status.

At college, Janna had no idea that she was interested in science. She was doing a philosophy major, but she was becoming increasingly frustrated in classes as it became clearer and clearer that 200 years on, people were still arguing about the meaning of Kant's writing and little else. In its semantic thoroughness, it can seem as if philosophy has a resistance to progress. One day there was a special lecture about quantum mechanics. This was when Janna fell in love with science. She realized that quantum mechanics was the same for everybody, and that whilst people may have different interpretations and theories, ultimately there were unavoidable truths about it, whatever you believed. Paradoxically her desire for something a bit more certain came from hearing something about a universe of probabilities.

Her thinking was, 'Oh my God, this is true in Bangladesh. This is true on Andromeda. This is just true. Once you spend the time learning it, it's yours. You can teach it to anyone else. Nobody's going back into Einstein's personal history to figure out what relativity is. Relativity is clear. And that moved me totally. It shifted me... it was shared, transcendently with all of us. And that's when I made the shift to mathematical physics, especially

358

the mathematical side, because, you know, maths is just true. That's unbelievably transcendent.'

Dropping circle stones on a sundial

The German sociologist Max Weber spoke of *Entzauberung* or de-magication – more neatly translated as disenchantment. He believed this was what science had done to the world. The world was now predictable; the gods and imps and trolls had been banished. It was an alienating place without transcendence.

Stuart Clark is an astronomer with a fierce love of prog rock. Growing up, his first favourite record was *Jeff Wayne's Musical Version of The War of the Worlds*, combining synthesizer bombast and extraterrestrial life contemplation. He has come to realize that the prog-rock ideals of big, ambitious musical compositions that have a core of truth and emotion in them, while letting the imagination go wild, could pretty much sum up cosmology. He wonders how valid the division is between our reactions to art and those towards science. We can all be moved by a piece of art – it is something we can feel. And we might also experience this when looking at the *Apollo 8* image of *Earthrise*. We can feel aesthetically that it is a truly beautiful thing, but then it hits us intellectually.

Stuart describes the change in the sky that we see when we allow our scientific curiosity access to it. 'We're not just looking at the night sky, populating it with gods and spirits and strange forces to have that kind of mystical wonder; we're actually realizing that these things are aesthetically beautiful – the universe, the night sky, planets, nature. It's all ecstatically beautiful in the same way that it is beautiful that a human being can create art. And yet we can also understand it at an intellectual level. And the two together are powerful.'

There is a serenity that we feel when we are under a dark night sky. The disenchantment that began with the scientific revolution in the seventeenth century, the transformation of nature into cold mechanical processes, is something Stuart believes has turned around, and that now we can appreciate both the beauty of something and how it works, and that both interact together.

Charles Darwin wrote in his journals when he travelled the world on HMS *Beagle*, experiencing sights that were beyond his imagination as he walked through the rainforests:

> The delight one experiences in such times bewilders the mind; if the eye attempts to follow the flight of a gaudy butter-fly, it is arrested by some strange tree or fruit; if watching an insect, one forgets it in the stranger flower it is crawling over; if turning to admire the splendour of the scenery, the individual character of the foreground fixes the attention. The mind is a chaos of delight, out of which a world of future & more quiet pleasure will arise.[9]

This is what I seek from both art and science: a chaos of delight. When I see someone explain bundles of particles colliding into each other at speeds near that of the speed of light, there are moments of transcendental wonder that are as impossible to define as the delight I feel when watching a performance by Nick Cave and the Bad Seeds. I am merely dizzily beamish that human imagination is capable of such things, and that this human imagination is driven by curiosity – a curiosity that can be expressed through both artistic and scientific creativity.

Watch a group of professors sharing their delight together about new research and new questions and you will see all the vim of Jackson Pollock wildly dripping paint onto a canvas. I believe it is vital that no one should leave school believing that science is a dispassionate pursuit where the imagination can't run wild. Science needs the human imagination, and once you have reached the outer limits, then you have to start testing where your imagination has taken you. Sometimes you will discover that it is all too improbable; but sometimes the utterly improbable turns out to be true, at least for now. It takes a lot of imagination to start working out the universe. Rather than scientific curiosity requiring you to suppress your imagination, it requires you to train your imagination, but still let it go mad in a field every now and again, whether that is a cornfield or a quantum field.

It is above all by the imagination that we achieve perception and compassion and hope.

URSULA K. LE GUIN

CHAPTER 12

So It Goes – Facing Up to the End of Everything

'We know how the Universe ends,' said the guide, 'and Earth has nothing to do with it, except that it gets wiped out, too.'

'How – how does the Universe end?' said Billy.

'We blow it up, experimenting with new fuels for our flying saucers. A Tralfamadorian test pilot presses a starter button, and the whole Universe disappears.' So it goes.

Kurt Vonnegut, *Slaughterhouse-Five*

I am always impatient to know how something ends. If my wife starts to watch a Scandinavian murder thriller, I do not have the patience to sit through ten episodes of subtitled twists about incest and betrayal. Within thirty-nine minutes I will go to my laptop and type in the title of the show and 'spoiler alert'. She looks at me furiously and I swear an oath of silence. One day I will make a Scandinavian murder series about a serial killer who murders people who publicize the spoilers of Scandinavian murder series. The universe is no different; I know there are still quite a few seasons and reboots to go before it all comes to an end, but I want to know what happens, because I know I won't be there at the end – almost nothing will. If you think that *Lost* had a disappointing series finale, wait until you hear about the cosmos.

Your death turns out to be a very parochial affair, almost whimsical, when you compare it to the seemingly inevitable end of absolutely everything. The universe's insistence on the necessity of death doesn't preclude it from having to follow that rule, too. Unlike its ability to break the speed of light, it is not currently believed that the universe will be able to escape what everything else is bound to face: extinction.

One day, *all of this* will end and, before it ends, for billions, perhaps trillions, of years, it may be very, very dull. You would end up praying for it to end with something as spectacular as a whimper, when you realize quite how boring those final years will be. It is a disquieting thought that one day the universe will be devoid of thoughts. At some time in the universe's future history, someone or something will have the final thought within the universe, their dying cogitation the very last cogitation of all. The universe will go on, but with nothing to look at it or question it. I think it is right to be disquieted by this, but it also reminds me of the absurdity that thoughts should be part of a universe in the first place.

My first encounter with the idea of the end of the universe was in Milliways, a restaurant designed by Douglas Adams, whose USP was that it was indeed situated at the end of the universe. I don't remember it disturbing me at the time. I was eleven years old, and the end of the universe seemed a long way off (it still does).

Why would I be disturbed by the universe ending when I was eleven? I was dealing with the more immediate daily terror of school bullies, and the mockery brought on by my inability to get over the wooden horse during physical education. My existential anxiety has always been tangible, even when physical fears have seemed to overwhelm the metaphysical fears. In fact I needed to obsess about the small picture, because if I looked up at the stars too long, wondering how long they would continue to shine, someone would tie my shoelaces together and I would end up flat-faced on the gravel.

Yet any fear about the end of the universe, as described by physicists, is possibly allayed by our inability to imagine nothing: the impossibility of picturing the obliteration of everything may be a stroke of luck.

When talking about the end of all existence with physicists, I find delight in the great big smiles on their faces. Sitting with three of them, Katie Mack, Brian Greene and Brian Cox, I see that their joy is not in the end of the universe, but in the possibility that we may be able to work out the end, as bleak as it might be. They run through possible destinies – a destiny that ultimately says there is no grand meaning, there are no more destinies for *anything* – and the mood is jovial, almost giddily optimistic. No one is evading the first issue of a universe of ever-increasing entropy, which is that whichever way it goes, the universe will be inhospitable for all living things for a very long time. There is far more ahead of us than behind us; 13.7 billion years is still cosmic toddler days, but what lies ahead has limited opportunities for beings like you and me.

There is another possibility, and that is that you are that Boltzmann brain and you are very close to the end of the universe: you, and only you. Nothing else exists, and your experience of this sentence is a remarkable illusion that occurs due to enormous periods of time and probability, but we'll get to that later on – or at least you will.

It is worth contemplating the rarity of complicated life in the universe at this very moment, from the evidence we have so far. If you are fearing your insignificance, think how significant it is that atoms have such a small window of opportunity to be part of complex structures – vast from a human timeframe, minuscule from a cosmic timescale. There is significance in the fact that every structure you see can exist. Look at the solid shapes around you, start to imagine them dissolving, the forces within them losing their grip, your reality thinning, the colours being turned down further and further, solidity being lost. As Carlos Frenk told me, each step further back in time creates a denser universe, and so each pace forward in time creates a universe that is a little less cohesive than yesterday. Things will fall apart; the centres of things really will not hold. The empty space of the universe will get much emptier, the features will become featureless, the points of interest will become pointless, so consider how lucky you are to be alive at the right time for looking at things.

There once was the steady-state theory, where the universe had no beginning and no end. It kept expanding, but new matter kept coming into existence, so the density always remained the same. This is one of the few scientific theories that was influenced by a horror-movie classic. Fred Hoyle went to see the unsettling portmanteau film *Dead of Night*, that film I mentioned earlier with demonic ventriloquist dummies. The story begins with an architect approaching a house with foreboding. He feels as if he has lived the experience before. Once inside, he meets the strangers, all of whom have a tale to tell, and his sense of déjà vu expands. It leads to a grisly conclusion, which is revealed to be a dream. Now awake, the architect goes off to his meeting of the day and approaches a house – the same house as we saw at the start of the film – and the story begins again. This inspired Hoyle to consider this to be the destiny of our universe: that it is on a cycle, with no beginning and no end.

Once the Big Bang model comes into play, though, starting from a singularity of infinite density and mass, future history changes. For those not prepared for the end being 'nigh' but with a pessimistic bent, a paper was published in 2010 suggesting that time would come to an end roughly at the same time as the Sun dies. This still gives you a few billion years. This was a consequence of the multiverse theory. The multiverse theory says that anything that can happen will happen, which makes calculating the probability of events tricky. It also offers an alibi: 'Sorry I got my predictions wrong, I was obviously thinking inside the wrong universe.' Theoretical physicist Professor Paul J. Steinhardt calls it the theory of anything. Not only does it predict that anything that can happen will happen, but that it will also happen an infinite number of times.

To try to avoid the prediction of the end occurring when the Sun ends and other such scenarios, mathematicians came up with an idea called 'geometric cutoffs', which means taking only a sample number of multiverses. According to theoretical physicist Raphael Bousso, geometrical cutoffs can't be used as mere mathematical tools 'that leave no imprint'. Bousso said that 'the same cutoff that gave you these nice and possibly correct predictions also predicts the end of time. If you use a cutoff to compute probabilities in eternal inflation, the cutoff itself becomes an event that can happen.' Stop mathematicians now! They'll get a god complex and destroy us all with their equations and paper logic.

More positive predictions see a universe existing in one trillion years, although there will be no new star formation or supernovae blasts. Should living, curious creatures exist in this time, the universe with its vastness and far greater distance between stars and planets will seem an isolated place. Distances will be so great that the kind of observations that have allowed scientists to gain an understanding of the birth and evolution of the universe today will be impossible.

What we can observe and measure now, to build the picture of our expanding universe, will be so far away that curious living creatures will be unaware of it. This will not be a terrestrial problem, as we will be out of the picture, due to the more parochial matter of the expansion of the Sun. Again this underlines a sense of significance. How can those civilizations that have been fortunate to live in this time of cosmological calculation and understanding on such a scale ensure that the information is preserved for future possible civilizations? And with them being unable to observe it themselves, would they even believe the records that were left via some form of intergalactic broadcast that would exist in perpetuity?

Perhaps this is why the cosmologists I talked to about the end of the universe are smiling, because they know they are the lucky ones, their life is timed just right to be able to start creating a picture of the future of the universe and understand its past. They are not happy about what is predicted, but they are happy that things such as this can be predicted… at least I hope that is the reason.

One of the comforts when we think about our own death is perhaps that memories of us may live on in the minds of others. So is there any comfort in a universe that progresses to a stage where there is no possibility of any memories, or experiences, as no one is around? What will be its point? Is that the point? It is hard to get to grips with the idea of our personal annihilation for those who see no prospect of an afterlife, but what of the end of all existence? It is one thing to think your life has been for nothing, but to think that all human endeavour has been for nothing…?

Paul Dirac is considered one of the greatest physicists of the twentieth century. Einstein described the Dirac equation as 'the most logically perfect presentation of quantum mechanics'. The Dirac equation brings together quantum mechanics and the special theory of relativity. It explains how particles like electrons behave when they move at speeds near that of the speed of light. Dirac was known for his precision, but not for his ebullience in social situations or his easy banter. Those mundane, but inevitable comments about 'the weather' that litter English small talk were of no use in engaging Dirac. If you said that it was cold out, he might reply, 'How cold?' He was not being facetious, he just liked to be exact in all things.

A dinner companion once commented that it was windy, and Dirac left the table, went to the door, peered outside and returned to confirm that the information was correct. As a scientist, mere anecdotal evidence of the windiness of a day was not enough for him. This precision found no room for a god or an afterlife. He was resigned to his imminent non-existence. Its inevitability did not bother him, but he was greatly disturbed by the end of everything. His afterlife was his work, its importance would live on; but what if there came a point when all the knowledge accrued by human beings was destroyed, when all the knowledge gathered by any and all curious, conscious creatures that have ever lived in the universe was annihilated? Then what was the point of it all?

Dirac wrote, 'In my case this article of faith is that the human race will continue to live for ever and will develop and progress without limit. This is an assumption I must make for my peace of mind. Living is worthwhile if one can contribute in some small way to this endless chain of progress.'[1] Even a bleakly rationalist thinker like Dirac could not countenance the idea that the chain might break or end.

Katie Mack is a smart and funny cosmologist. She has good advice for any cosmologist veering towards narcissism: 'Being the centre of your own universe may sound appealing until you realize everything is trying to get away from you as fast as possible.' Her discipline means that she can't avoid confronting how powerful the cosmos is, and how powerless we humans are. When she was in grad school, reading about supermassive black holes and the centres of galaxies, she started to get a sense of our inconsequentiality. 'Supermassive black holes can create jets of radiation that stretch for thousands of light years, and they can tear apart stars and all of this. Those acts of extreme violence are happening in the universe all the time, and it's just luck that it hasn't happened to us on this planet. We're here at the moment, but various things could happen in the cosmos that would destroy us,' she explains.

She tells me about a morning coffee gathering of an astronomy group, where they talk about the papers that have come out the night before. She tells me about one of the papers that cropped up over the espresso. As she explained, I did not really understand what it all meant initially. It was about a white dwarf star that had what was described as contamination and spectrum. Katie explained, 'This contamination was that the star had spectral signatures of heavier elements.' The spectral signature shows what chemical elements there are in an atmosphere. The spectral signature of this star was unusual, in that it showed the presence of silicone and other things that would usually exist on a planet, and not on a star. From what was discussed as the caffeine kicked in, it appeared that this star had torn apart a planet, and also that there were stars like it all over the universe.

As Katie thought about this, she had a moment of utter horror that a whole planet had been eaten by a star and everybody else in the room was pretty blasé about it. It was not remarkable to them, because these things happen in a universe like ours. Katie thought of the swallowed planet. Did anyone live on it? Had a civilization been destroyed by a hungry star? What if there were intelligent beings, and the star threw off its underlayers and became a white dwarf and the planet fell in? To her, it was a vision of extreme violence at a size that we rarely contemplate.

Like an Eastern European film melodrama about a widowed coffin maker pestered by gout, Katie's lectures don't start happy. Observing stars in the universe, we know the destiny of our star. It is going to expand into a red giant. The oceans of our planet will boil away, and eventually Mercury and Venus will spiral into the Sun. Then there will be a white dwarf star in the Milky Way galaxy with a little bit of contamination in its spectrum, and that contamination will be us. When we see the signatures of these extreme events in other places in the universe, it could very possibly be our own fate that we're looking at.

Katie tells me that when she talks about the heat-death of the universe – currently the theory for the end of the universe most favoured by cosmologists, the one that sees everything fading away – there is often a measurable sense of sadness in the room. This is a reaction above individual ego. The individual ego might mourn its own demise, but not really care what happens after that; after all, it won't be around to witness it. This is physics to which people have an emotional response. Like a doctor with a negative prognosis, she gives the news in a careful and considered manner. Katie has had people in the front row putting their heads in their hands, while other colleagues have had people crying during their lectures.

She understands that it can be very hard for people to come to terms with the idea that the universe will probably fade away over a long, slow darkness. It's interesting, but it is highly unsettling. There are other scenarios – some of them very dramatic, some very scary – and Katie will play that up because she think it's a more exciting story; but when it comes down to it, she really thinks, 'The likelihood is that everything will kind of fade to black and it'll be this dark, empty universe for ever... Even other physicists often don't want to accept it, you know, and I've certainly talked to many physicists who think maybe there's something else, there's some other answer, because it is kind of a sad and a bleak sort of future.'

Katie takes great care introducing the heat-death, especially if there are children at the lecture. She explains that kids don't really make a distinction between 'ends tomorrow' and 'doesn't last for ever'. Just because it's going to be a really, really long time from now and they don't have to worry about it, that doesn't mean they won't worry about it, as children can find any kind of ending quite upsetting. It makes her focus on the idea of making meaning, because there is no other meaning to be found, apart from what you can create in your own mind.

Then, she wonders, can there be meaning and purpose if there's no record of it in the end? Katie feels that we are so predisposed to think that the final outcome is what retroactively justifies everything else. If the last episode of the TV series that you love does terrible things to the main character you empathized with, you come to hate the series, because you now know that it was always heading towards that terrible ending. Katie thinks it's very hard for us to separate purpose and meaning from the ultimate consequence. 'If there is no ultimate consequence – if everything that happens to us is, in some sense, inconsequential, if it doesn't have a lasting impact on something larger – it's very, very hard to see how we say it has meaning.'

Kate has spent much of her recent career pondering the end of the universe and lecturing about it. On her own death, she believes that even if she is completely forgotten in the future and nobody has any idea she ever lived, there is still some sense in which what she did in life matters, because she had an effect on the people around her – maybe she inspired somebody who did something interesting. She believes this is something we definitely do get from physics: 'that everything affects everything else and you can't have a totally inconsequential thing in a system'. Everything does build on everything before it, and there are these very complicated relationships between things, 'so my personal death does not erase me and my purpose'.

She thinks spending too much time thinking about the ultimate destruction of everything is definitely not healthy, though one upside is that you could also think, 'Don't worry about the mess you leave behind; it will ultimately be cleared up by the destruction of everything', and you could either sink into a life of inaction or embrace the sort of nihilism favoured by the current crop of politicians. For me, I think the pointlessness leads to wanting to create a localized sense of meaning for my time of existence.

Although Katie wrote a book on the end of the universe, which ratcheted up her existential anxiety, it also gave her some useful perspective. A year before writing it, her grandmother passed away. In her last years, dementia had made communication harder. For a while Katie was sad that she was gone and there was very little to remember her by. Then she had this moment of thinking, 'We are all in this situation. None of us will have a lasting impact.' While her grandmother was here, she had an important and loving impact on Katie, who thinks that's got to be enough.

Katie concludes, 'Cosmology is the study of the evolution of the universe and it is not a happy ending.' But what is a happy ending?

British astrophysicist Chris Lintott agrees that the universe is going to get more empty and boring, but he is anxious about the idea that we live in a special time or place. He looks up and sees an interesting era, in which both galaxy mergers and gin-and-tonics exist, and this can make him feel queasy. On a good day he accepts the queasy feeling as a scientific one because he's imbued with the Copernican principle, but it still makes him uneasy that we are living in a special time of the universe.

'People have written all sorts of bizarre and interesting papers on how to reconcile that – whether it's just a bias because we're observers, so we need to live on, or whatever else.' But Chris also tells me that he does try to skip some of those thoughts and feelings. Sometimes, at the end of a talk, he can hear the audience muttering, 'I didn't really want to think about that. I don't know how to think about that. I don't know how to think about the fact that there are more stars dying now than being born in the universe.'

Although we may wish to believe in the stout objectivity of scientists, what draws you to a scenario may be that it is more personally satisfying. This is not about dishonesty, but what you focus on will be connected to how you may feel about the world.

Death drives and end times

I am intrigued by our different reactions to the end. Tell someone they will die in twenty months and everything changes. Tell them twenty years and they may be reasonably nonchalant. Tell them everything is destroyed in trillions of years and there is a knot in the stomach again.

I asked a Freudian friend, therapist and lecturer Josh Cohen, what he thought about our reactions to the termination of all. He told me that Freud talks about two forms of drive: a personal drive towards one's own pleasure and gratification, and an impersonal drive towards self-reproduction and continuity of the species. He sees the self-preservation – the care that we take with our own life – as our alertness to the broad issues of health and safety. Freud views that as evidence of the species' own will to continue.

Josh thinks one of the things that makes our own death tolerable is knowing that others will continue after us – that the world will live on after you and me. We will dissolve into it in a way that leaves some kind of mark, some kind of trace. The fact that we live with the history of previous generations also helps us accumulate a sense of the continuity of the generations. If all that is annihilated with the death of the universe, Josh sees us being left with nothing that appears to justify existence. When he says it like that, I find it hard to work out why I am not bothered by this outcome. Perhaps I just have a short-term imagination, or perhaps I can accept that I simply get a brief flash of existence and then I am gone, and that somehow allows me not to be too worried that everything goes in the end. The end of everything is merely another absurdity and makes all our arrogance and enterprise and searching for meaning seem even funnier.

What's it all for? Nothing. Oh well, at least it passed the time.

Although the equations that are required to predict the various possible annihilations may be beyond my capabilities, I think I am beginning to understand why every physicist is smiling. I think I find it quite funny, too. You have to confront the ultimate meaninglessness, when confronting the end of the universe. 'Why bother doing anything at all if it is all for nothing?' But if you don't do anything, it would be incredibly dull. Sure, Ozymandias, your empire has crumbled to dust, but there was some fun in the act of building it (and probably some appallingly lax health-and-safety too).

If there is no point in any of it, and the ultimate judgement due to the laws of physics is total destruction after all, you might as well work out how to create joy in yourself and others. If you now know that the vast majority of the time of the existence of the universe offers no opportunity for anything at all, apart from the scant possibility of that cosmic grand piano coming together, then focus on what can seem like good fortune. Enjoy the journey, even if it leads nowhere. We have evolved to believe there is a destination, and yet so often when we get to the place we thought we wanted to be, we are soon bored or dissatisfied.

Do not be sad that the universe will spend much of its life being utterly boring; be glad that you are not having to put up with such tedium, and try to avoid boredom yourself. Take the advice of all those ISS astronauts: make sure you look out of the window.

We are all creatures of the stars.

DORIS LESSING

Afterword

There is nothing new under the sun,
but there are new suns.

Octavia E. Butler

I t feels a bit brutal to conclude a book with the death of all things and of all the thoughts that have ever existed, so unlike at the death of the universe, I thought I could add an afterword of sorts.

I am writing this on 25 March 2021, a day when I am directly benefiting from the well-aimed curiosity of human beings. There is a dull throb in my upper left arm and I feel a bit groggy. I don't know if I feel groggy from the vaccination against Covid-19 or if I am simply groggy because I am fifty-two and middle age leads to bouts of irrational grogginess. Racing around my body is a scientific success that may mean I am allowed out of my attic and can stand in front of real-life audiences before the end of the year – like many people, getting back to the things I love doing.

Before the pandemic I had spent the year travelling the world and performing in front of hundreds of thousands of people. During the pandemic, if I didn't make too much noise, my family would sometimes let me out of the attic and down the steps for some biscuits. We have all faced abrupt changes, and my sometimes shaky grip on sanity has been tested over this year, as I am sure yours has, too.

I really don't like staying still. Actually it is not a matter of not liking – I just can't. I need to jump into busyness. I need to have projects all around me. In the early days of the pandemic, when all that came to a stop, I was struck by a sledgehammer headache, unlike anything I had experienced before. I had an inkling of what was causing this and started typing; 3,000 words later, the headache was gone. My inactivity had caused a build-up of words in my skull, which felt like physical pressure. I typed them out of my head, tried to make sense of something by putting it into sentences and I felt better. Perhaps if I had blown my nose after eating too much Alphabetti spaghetti there would have been a similar effect. In a year of possible stagnation, the importance of being interested and curious became even more urgent. I need to keep feeding my brain.

I hope in this book I have explained why I believe it is so important to be interested about our world and the universe that surrounds it. I despise boredom and, when it does hit me, I know it is my fault. All those times that I have read about the human brain being the most complex known thing in the universe, then I find myself allowing mine to be bored, and I think how boring that must mean I am. I think about all the pain and risk that human mothers go through during childbirth, because of the great big skulls that hold our great big brains, and it annoys me that I might have reached a point where I can't think of anything to do with my brain.

Yet I don't need anything more than a brain to think myself out of such boredom – to stare out of a window and focus on a cow in a field or the shape of a cloud, or just look down at my fingers and start thinking of all those mutations that led to a hand with a thumb, and the advantages that gives us for hitch-hiking or ringing a bicycle bell or making tools, though when it comes to DIY, my thumb's main purpose seems to be to remind me how heavy a hammer is.

The more time we spend being interested in things, the fewer opportunities there are for boring thoughts. The limit of our curiosity is the limit of our world. The more wonder that we have to wonder about, the more places we can go in our minds. It gives us more rooms, more landscapes, more planets and stars.

I have interviewed more than 100 people while writing this book, and I cannot imagine any of them being bored. They are all constantly interested and interesting. When we record *The Infinite Monkey Cage* it always takes longer than we plan for because, once people start talking, the ideas that fly around generate new ideas and new questions. Rarely do we have guests who merely want to present what they already know. They are always wrestling with their old ideas and grasping for new ones. On many recordings we don't get past the first planned question, because other questions just keep spawning.

For all the practical uses of our sense of curiosity (which my aching upper arm is testament to), there is also the fact that everything has the potential to be interesting, to be provocative. Ideas have more to bounce off. Things connect. Webs are made.

The mundane has the potential to become fabulous. The fabulous has the potential to become comprehensible. The inscrutable is beguiling. Change is possible. Everything contains a story.

Things easily dismissed as ugly or dull are revealed to have magnificence. Some of my favourite episodes of *The Infinite Monkey Cage* are when we deal with something that people don't normally spend much time thinking about, but which they might experience every day. Our episode about flies caused great excitement. Dr Erica McAlister from London's Natural History Museum must be the most ebullient proselytizer about flies. It doesn't mean their buzzing doesn't still annoy me, or that I now like seeing one vomiting on my Chelsea bun, but to me flies are now so much more than the nuisance they once were.

For me, the importance of being interested – beyond all the remarkable achievements, the cures and the communication, the dismantling of destructive ideologies, the solutions to our most pressing problems, the grandeur in these views of life – is that happiness and purpose can be found in perpetual curiosity.

By about age twelve, I would prefer to stay up and watch the stars than go to sleep. I started learning. I started going to the library and reading. But it was initially just watching the stars from my bedroom that I really did. There was just nothing as interesting in my life as watching the stars every night.

VERA RUBIN

Notes

Introduction: The Stars Your Destination

1. 'The Poet and the City', *The Dyer's Hand* (1963)
2. *Has Religion Made Useful Contributions to Civilization?* (1930), although I first read it in Brian Greene's excellent *Until the End of Time* (2020)

1 Scepticism – From the Maelstrom of Knowledge into the Labyrinth of Doubt

1. 'Whoever sows sparingly will reap sparingly, and whoever sows generously will also reap generously'
2. https://hayleyisaghost.co.uk/horror-skeptic-movement/
3. According to the author Michael Meek, 'Shortly before he left the BBC in 1990 he experienced a metaphysical epiphany in a newsagent's on the Isle of Wight.' *London Review of Books*, 22 October 2020
4. *The Demon-Haunted World* by Carl Sagan and Ann Druyan (1995)
5. This is one of many articles on how this has worked: https://www.theguardian.com/environment/2019/oct/10/vested-interests-public-against-climate-science-fossil-fuel-lobby. I would also recommend the book *Merchants of Doubt* by Erik M. Conway and Naomi Oreskes (2010).

2 Is God on Holiday? – Are There Still Enough Gaps for a God?

1. My friend Grace Petrie wrote a song inspired by this letter and I highly recommend that you listen to it. She lives at: www.gracepetrie.com
2. https://www.theguardian.com/commentisfree/2016/sep/06/david-jenkins-bishop-durham-biblical-facts-fire-york-minster
3. https://www.bbc.co.uk/sounds/play/p00949ct
4. https://www.quantamagazine.org/entanglement-made-simple-20160428/

3 Armchair Time-Travel – Putting Out Your Beach Blanket on the Sands of Time

1. Written by Harlan Ellison
2. https://quoteinvestigator.com/2019/07/06/time/
3. Please see 'City of Death' – a *Doctor Who* story co-written by Douglas Adams
4. St Bede, *Ecclesiastical History of the English People*, Book 2, Chapter 13
5. http://www.bbc.com/earth/story/20161012-the-strange-origin-of-scotlands-stone-circles
6. https://www.nytimes.com/1987/03/12/us/whale-fossils-high-in-andes-show-how-mountains-rose-from-sea.html
7. https://www.nationalgeographic.com/science/2018/08/news-happens-plate-tectonics-end-earth-mountains-volcanoes-geology/
8. https://londonhollywood.wordpress.com/2016/09/26/the-alan-moore-jerusalem-interview-tapes-3-stop-pushing-the-wheelbarrows/
9. This idea is brilliantly explored by Ted Chiang in *Story of Your Life*, impressively adapted into the film *Arrival.* I recommend both heartily.

10. *The Rubáiyát of Omar Khayyám*, translated by Edward Fitzgerald (1859)

4 Big, Isn't It? – On Coping with the Size of the Universe

1. https://cosmosmagazine.com/space/how-big-is-the-universe/
2. *Alexandria 3: The Journal of the Western Cosmological Traditions*, edited by David Fideler (1995)

5 Escape Velocity – On Looking Back at the Planet from a Height

1. http://www.jgballard.ca/media/1996_seconds_magazine.html
2. 'A Reflection: Riders on Earth Together, Brothers in Eternal Cold', *The New York Times*, 25 December 1968
3. Presented at the Lindisfarne Conference in 1974, and first published in *Earth's Answer*, edited by Michael Katz, William P. Marsh et al. (1977)
4. First said on *Fresh Air*, National Public Radio, 31 August 1993
5. https://www.theguardian.com/lifeandstyle/2016/apr/18/blast-off-why-has-astronaut-helen-sharman-been-written-out-of-history
6. http://www.jgballard.ca/media/1996_seconds_magazine.html
7. 'Astronaut Special', *The Infinite Monkey Cage*, 11 July 2017; https://www.bbc.co.uk/programmes/b08x8y1g

6 Why Aren't They Here? Or Are They...? – On Waiting for Our Alien Saviours

1. This is quoted in many publications and is always attributed to Arthur C. Clarke, though I can't find the original source. The useful website Quote Investigator has found the earliest near-match attributed to Stanley Kubrick, quoting an unnamed

scientist who is presumed to be Arthur C. Clarke as they worked so closely on *2001: A Space Odyssey.*

2. In *Star Trek*, 'The Devil in the Dark', Season 1, Episode 25

7 Swinging from the Family Tree – Inviting Yeast to the Family Reunion

1. Here is a piece he wrote in 2004: https://chem.tufts.edu/answersinscience/MillerID-Collapse.htm. Obviously you can read more up-to-date arguments about the evolution of the flagellum because, like everything in science, the answers continue to develop

2. https://peped.org/philosophicalinvestigations/extract-1-the-flagellum-unspun/

3. From Robert Ardrey's introduction to *The Soul of the Ape* (1969)

4. https://www.dbnl.org/tekst/mara002vers01_01/mara002vers01_01_0191.php

5. https://www.chicagotribune.com/news/ct-xpm-1990-08-01-9003040522-story.html

6. https://www.masterclass.com/articles/chimpanzee-intelligence-with-dr-jane-goodall

7. From the documentary *Jane*, directed by Brett Morgan (1987)

8. https://www.theguardian.com/environment/2014/jun/08/the-dolphin-who-loved-me

9. http://content.time.com/time/health/article/0,8599,1973486,00.html

10. https://www.lbhf.gov.uk/sites/default/files/section_attachments/the_economics_of_biophilia_-_why_designing_with_nature_in_mind_makes_financial_sense.pdf

11. https://www.americanscientist.org/article/perceptual-pleasure-and-the-brain

12. Ibid.

13. https://pdfs.semanticscholar.org/bfa5/2d19c047b0c109ee5db

4b11d1c4dc0f3dc74.pdf?_ga=2.160479034.649629674.
1596273888-98647138.1596273888

14. Frances E. Kuo and William C. Sullivan, 'Aggression and Violence in the Inner City: Effects of Environment via Mental Fatigue', *Environment and Behavior*, 33 (4), 1 July 2001, pp.543–71

15. *Evolutionary Perspectives on Environmental Problems*, edited by Iver Mysterud (2007)

16. *On the Origin of Species*, 6th edition (1872)

8 The Mind Is a Chaos of Delight – On the Matter of Grey Matter

1. From the introduction to *The Origin of Consciousness in the Breakdown of the Bicameral Mind* (1976)

2. *An Alchemy of Mind: The Marvel and Mystery of the Brain* (2005)

3. 'The Human Brain', *The Infinite Monkey Cage*, 27 August 2020; https://www.bbc.co.uk/programmes/p08kn47j

4. https://www.insidehook.com/article/history/winston-churchill-historic-wwii-speech

5. http://staff.washington.edu/eloftus/Articles/Bulllpsychologistpdf HIGHRES01.pdf

6. Ibid.

7. https://www.sciencedaily.com/releases/2003/02/030217115223.htm

9 Reality, What a Concept – Can Anything Be What It Seems?

1. The title of Robin Williams's debut stand-up album

10 Imagining There's No Heaven – On Being Finite

1. *An Autobiography and Past Forgetting* by Peter Cushing (1986)

2. https://www.theguardian.com/society/gallery/2008/mar/31/lifebeforedeath

3. https://royalsocietypublishing.org/doi/pdf/10.1098/rstb. 2001.0885
4. *Waiting for Godot* (1953)

11 More Important than Knowledge – On the Necessity of Imagination

1. *Coming of Age in Samoa* by Margaret Mead (1928)
2. Though this quotation is often attributed to W. B. Yeats, there seems to be much debate as to who originally said it, and current thinking is that it is based on the words of the Platonist philosopher Plutarch.
3. From 'The Imaginative Mind in Science', *The Visionary Eye* (1978)
4. *Unended Quest: An Intellectual Autobiography* (1976)
5. *The Universe in a Glass of Wine* (2011) and *How to Start a Feynman* (2011). I recommend that you hunt for both.
6. https://www.brainpickings.org/2018/01/09/richard-feynman-poetry-science/
7. 'Darwin and the Meaning of Flowers', *The River of Consciousness* (2017)
8. https://edition.cnn.com/style/article/olafur-eliasson-in-real-life/index.html
9. A journal quote taken from Randal Keynes's *Annie's Box* (2001)

12 So It Goes – Facing Up to the End of Everything

1. *The Strangest Man: The Hidden Life of Paul Dirac, Quantum Genius* by Graham Farmelo (2009)

Acknowledgements

A huge thank you to everyone who gave me their time when I was writing this book.

As well as thanks to all those who have appeared in the book, there were a vast number of interviews that I did not use, not because they were not fascinating but because my publishers quite rightly didn't want a book that was as long as one of those great big biographies of Stalin that give you sciatica. So thanks to those you did not see, but who helped shape my mind as I battled with the universe who included Lisa Dwan, Karen Rodham, Paul Zenon, Janina Ramirez, Natalie Kay-Thatcher, the Bishop of Leeds aka Nick Baines, Stephen Volk, Bob Fischer, Eric Idle, Julia Hamer, Andrew Copson, Cath Loveday, David McAlmont, Ray Tallis, Ben Bailey Smith, Uta Frith, Chris Frith, Sarah Bakewell, Danny Hills, James Kennedy, Susan Rogers, Angie Hobbs. I will write the book I thought I was writing back in March 2020 and put you all in that.

Thank you to my friends Helen Czerski (listen to our regular *Sunday Science* Q&A), Johnny Mains (read our hour anthologies) and Carolyn Wilson (see our movie *Razzle Dazzle: A Journey Into Dance*), Allan Lear (a Gallifreyan improviser), Toby Evetts (my ancient LA friend) and Lee Randall (one of my favourite book festival interviewers) who read this book at various stages and gave me very useful feedback; much of the blame must be theirs.

To Carlos Frenk who was my regular Wednesday afternoon conversation above and beyond for the writing of this book.'

To Matthew Cobb who takes my frequent phone calls where I ask him, 'What's all that about then?' Similarly, to Carl Cooper, who has always been a joy to discuss ideas with.

I was fortunate to spend the year before lockdown traveling the world and so a huge thank you to Adam Scott and the brilliant crew, hopefully by the time you read this we are all back on the road again.

To both the Mikes for their editing skills (yup, it takes two to cut me down to size).

To the encouraging voices on social media (this is not sarcastic, there are plenty of genuinely lovely people who I communicate with there and they more than make up for the aggressive and negative voices).

Thank you to Trent and Melinda Burton for all the work we do together on Cosmic Shambles, which has also played its part in shaping this book and my mind, and to Josie Long, my frequent online companion during lockdown.

My *Monkey Cage* companions – Sasha, Maria and Brian.

And obviously to Nicki and Archie, who have had no escape from me during lockdown.

And I'll finish my thank yous for two people who will not be able to read them. Thank you to Gina Ryan and Stan Vernon, both of whom died in 2020. They were good friends on the comedy circuit and, as so often, life changed and I saw very little of them, but memories of strange and wonderful gigs remain.

Now who did I forget? I will leave some space here for you to add your name just in case.